W0113394

The Poetry of Physics

The Poetry of Physics explores the intersection of science and art, blending the intricate beauty of physics with the evocative power of poetry. This unique work takes readers on a journey through the physical world, from the delicate patterns of living organisms to the vast reaches of the cosmos.

Structured in four sections – living physics, environmental physics, celestial physics, and a guide on writing your own poems – this book offers both scientific insights and poetic reflections, providing a richer understanding of both fields. The final section provides practical guidance on crafting your own physics-inspired poetry, encouraging active participation in this tradition of blending scientific and artistic inquiry.

Ideal for those who appreciate both science and the arts, whether they are physicists, aspiring poets, or curious minds seeking to explore the world and our place within it.

Sam Illingworth is a Professor of Creative Pedagogies at Edinburgh Napier University, specialising in the intersection of science and the arts. With a PhD in Atmospheric Physics, he is an award-winning science communicator and poet, and founder of *Consilience*, the world's first peer-reviewed science poetry journal. Discover more about his work at www.samillingworth.com.

The Poetry of Physics
From a Quark to a Quasar

Sam Illingworth

CRC Press
Taylor & Francis Group
Boca Raton London New York

CRC Press is an imprint of the
Taylor & Francis Group, an **informa** business

Designed cover image: NASA, NIRCam

First edition published 2025
by CRC Press
2385 NW Executive Center Drive, Suite 320, Boca Raton FL 33431

and by CRC Press
4 Park Square, Milton Park, Abingdon, Oxon, OX14 4RN

CRC Press is an imprint of Taylor & Francis Group, LLC

ISBN: 978-1-032-84733-7 (hbk)
ISBN: 978-1-032-84731-3 (pbk)
ISBN: 978-1-003-51468-8 (ebk)

DOI: 10.1201/9781003514688

Typeset in Sabon
by SPi Technologies India Pvt Ltd (Straive)

For Becky, Cora, and Grace – poetically perfect in every way

Contents

Acknowledgements

I extend sincere thanks to the readers of my blog and the listeners of my podcast, 'The Poetry of Science'. Your engagement has been a wellspring of inspiration. I also wish to acknowledge my friends, family, colleagues, and students for their patience with my poetic explorations and their unwavering support.

Gratitude is due to the wider academic communities who have both embraced and scrutinised the merging of poetry and physics. These interactions have been instrumental in refining my craft as a writer, critic, and poet.

A special mention goes to the *Consilience* team, whose exceptional work in integrating poetry with science has been invaluable. Particular thanks to Mirjam Mahler, Allegra Biava, Dani Salvadori, and Kayla-Jane Barrie for their constructive and insightful comments on an earlier draft. Any errors that remain are solely my responsibility.

Chapter 1

Introduction

WELCOME

Scientific research findings are predominantly shared through scientific journals, where a concise abstract summarises the study. Journals typically mandate a word-limited abstract during submission, serving to succinctly encapsulate the research. These abstracts, while varying in specifics across journals, aim to offer a brief overview, enabling readers to gauge the study's relevance to their interests. However, these abstracts often use technical language that can alienate non-specialist readers, potentially keeping the research from those who might find it valuable or intriguing.

To bridge this gap, in 2014 I initiated a weekly blog titled 'The Poetry of Science' (https://thepoetryofscience.scienceblog.com/), aiming to introduce scientific discoveries to broader audiences through poetry. Initially, my readership was modest, consisting mainly of family and a few erratic bots, tallying 30–40 monthly readers. Yet, as interest in the blog grew, I was inspired to explore an unconventional question through my own research: could poetry effectively stand in for traditional scientific abstracts?

This hypothesis was examined with a group of scientists, chosen because any alternative to traditional abstracts must remain informative and engaging for this primary audience. In my study (Illingworth, 2016), scientists were divided into two groups to assess responses to a scientific abstract versus a poem I had composed, both based on the same research regarding climate change's impact on Canadian glaciers (Clarke et al., 2015).

DOI: 10.1201/9781003514688-1

Here is the poem that I wrote:

A glacial retreat

In Canada a study found,
How glaciers melt in the West.
The shrinkage is beyond profound,
With seventy per cent at best;
If we ignore the Earth's request
Then ninety-five per cent will go.
New barren lands will not be dressed,
With climate change too warm for snow,
The alpine streams and sapphire lakes they too will go

Quantitative analysis revealed that although a scientific audience considered a poetic rendition of a scientific abstract equally engaging and inspiring as the original text, they deemed it less accessible. Nonetheless, subsequent qualitative evaluation indicated that the poem effectively communicated a meaning akin to that of the original abstract. This outcome suggested that, at least for scientific audiences, poetry might not fully replace traditional abstracts. Instead of discouraging me, these insights prompted a re-evaluation of my poems' objectives, steering me towards enhancing their role as complementary narratives that encourage the reader to view the research through a new lens.

Following this study, I started to write short explanations of the scientific research to accompany my poems; these were aimed at a non-specialist audience, providing them with an overview of the study and its findings. Writing these short explanations allowed me to be less literal in my poetic interpretations of the scientific research; gone now (mainly) were the forced rhymes for 'thermonuclear' or 'photosynthesis'. My poems no longer aimed to replace the traditional scientific abstract, but rather to enhance it by offering the reader an alternative way of considering what this research might mean for both their own lives and the wider society.

The poems that feature in this book started their life in my blog, although they appear here in an edited (and improved) form, prefaced by a brief overview of the research that inspired them. I have also included references for the original scientific research articles, so that you can read them for yourselves and find out more about the incredible, life-affirming science that they contain.

So why *The poetry of physics?* As a lapsed physicist (see e.g. Illingworth et al., 2009, 2011; Ryder et al., 2010) I still find myself drawn to fundamental physics research, which is perhaps one of the reasons why a lot of my poetry is situated in this branch of science. But there is more to it than that. Physics and poetry, at their core, are both forms of exploration and expression. They each ask us to consider not just the 'what' of the world, but the 'why' and the 'how'. Physics seeks to answer these questions through experiments and equations, offering us a framework to understand the universe's mechanics. Poetry, in contrast, uses language, emotion, and imagery to explore similar questions, providing insight into the human experience within that universe.

LIGHT IS A BRIDGE BETWEEN PHYSICS AND POETRY

The beauty of physics lies in its ability to explain the vast and the minute, from the galaxies swirling in space to the particles that dance within atoms. These explanations do not strip the world of its wonder; rather, they enhance our appreciation for its complexity and intricacy. Similarly, poetry captures moments of clarity and insight, distilling complex emotions and ideas into words that resonate with our deepest selves.

The second law of thermodynamics, which highlights the inevitable increase of entropy, resonates with poetry's reflections on change, loss, and the flow of time. This parallel draws our attention to the ephemeral beauty of the world, underscoring the value of savouring moments of order and connection. Similarly, quantum mechanics challenges our conventional understandings of reality, urging us to accept uncertainty and recognise the limits of our knowledge. Poetry mirrors this sentiment, frequently navigating the realm of ambiguity and the interstices of words and meanings, inviting us to take solace in the unknown.

In a comparable vein, the concept of light serves as a bridge between physics and poetry. From a scientific perspective, light is at once a wave, a particle, and an enigma. Within the realm of poetry, light symbolises hope, insight, and the unveiling of truths that lay concealed in the shadows. Furthermore, the exploration of space, which has transitioned from the purview of deities and legends to a subject of scientific investigation, continues to fuel both awe and inspiration. The infinite expanse of the cosmos, targeted by physics for measurement and comprehension, equally acts as a backdrop for the poetic imagination, reminding us of our insignificance and, simultaneously, our capacity for magnificence.

The poems in this book are categorised into three themes:

- Part 1 – living physics, focusing on physics as it relates to humans and wildlife.
- Part 2 – environmental physics, which deals with physics in the context of the environment.
- Part 3 – celestial physics, covering the realms of stars and the cosmos.

This structure aims to explore the vast scope of physics, from the quantum level to the expanses of space, also reflecting my personal journey in these areas, shaped by my background first in space science and later in atmospheric physics.

This book does not seek to simplify physics into poetry or to complicate poetry with physics. Instead, it aims to show that the appreciation of one can enrich the understanding of the other. By looking at the universe through the dual lenses of science and art, we can find new ways to appreciate its beauty and its mystery. There has been a strong history of physicists writing poetry, from James Clerk Maxwell to Jocelyn Bell Burnell (Illingworth, 2021), illustrating that the interplay between these disciplines is not just possible but profoundly enriching. This tradition underscores the idea that scientific inquiry and poetic expression are both fundamental to the human experience of seeking knowledge and understanding (Illingworth, 2022). The blending of these fields invites us to explore the universe with a sense of wonder and curiosity, bridging the gap between the empirical and the existential, and encouraging a deeper engagement with the world around us.

In reading this book, I hope that you are inspired to find out more about the physics that is featured and that you might use my poetry as a way to introduce the topic to your friends, family, collaborators, and students. I also hope that you will use this to write physics poetry of your own. To facilitate this, the final chapter in this book presents a guide for how you might start to create work of your own and, in doing so, find your own rhythm within the physics of our universe.

REFERENCES

Clarke, G.K., Jarosch, A.H., Anslow, F.S., Radić, V. and Menounos, B., 2015. Projected deglaciation of western Canada in the twenty-first century. *Nature Geoscience*, 8(5), pp. 372–377. https://doi.org/10.1038/ngeo2407

Illingworth, S., 2016. Are scientific abstracts written in poetic verse an effective representation of the underlying research? *F1000Research*, 5, Article 91. https://doi.org/10.12688%2Ff1000research.7783.3

Illingworth, S., 2021. *A sonnet to science*. Manchester: Manchester University Press.

Illingworth, S., 2022. *Science communication through poetry*. Cham: Springer.

Illingworth, S.M., Remedios, J.J. and Parker, R.J., 2009. Intercomparison of integrated IASI and AATSR calibrated radiances at 11 and 12 μm. *Atmospheric Chemistry and Physics*, 9(18), pp. 6677–6683. https://doi.org/10.5194/acp-9-6677-2009

Illingworth, S.M., Remedios, J.J., Boesch, H., Ho, S.P., Edwards, D.P., Palmer, P.I. and Gonzi, S., 2011. A comparison of OEM CO retrievals from the IASI and MOPITT instruments. *Atmospheric Measurement Techniques*, 4(5), pp. 775–793. https://doi.org/10.5194/amt-4-775-2011

Ryder, S.D., Illingworth, S.M., Sharp, R.G. and Farage, C.L., 2010. The nuclear ring in the barred spiral galaxy IC 4933. *Publications of the Astronomical Society of Australia*, 27(1), pp. 56–63. https://doi.org/10.1071/AS09016

Chapter 2

Living physics

INTRODUCTION

In 'Living Physics', we explore the interplay between the living world and the constant laws that shape our universe. This segment reveals how we comprehend the workings of the world, employing both scientific insight and poetic expression to cast light on hidden realms and articulate the cadences of life.

Our journey commences with the rudimentary rhythms of human innovation, observing how primitive instruments cradled life but also captured melodies echoing both antiquity and the contemporary.

Advancing further, the collection confronts the contemporary trials facing the natural world and our efforts to protect it. We consider advanced techniques that unveil previously concealed wonders and methods that enable us to monitor the great roamers of both land and deep from the heavens.

The verses proceed to ruminate on the vibrant activity of the minute, where masses of minuscule beings generate an aura barely within our grasp. From the ocean's depths to the atmosphere's vast expanse, we detect vestiges of existence vibrating with the Earth's pulse.

Living physics beckons you to attune your ears, to gaze intently, and to resonate with the world in novel perceptions. It extends an invitation to realise that within each existence, no matter its size, lies a narrative of physics awaiting revelation.

The ordering of these poetic works follows a deliberate trajectory, threading from the ancient to the contemporary, aligning pieces that not only stand alone but also connect, casting a continuous narrative that flows from our origin to our present, from the micro to the macro. This arrangement provides coherence and context, allowing

DOI: 10.1201/9781003514688-2

us to perceive the unbroken thread of physics that weaves through the tapestry of life.

RECREATING ANCIENT MUSIC

The science

Nestled in the heart of the French Pyrenees, the Marsoulas cave, adorned with ancient wall art, serves as a silent guardian of history. Here, researchers discovered a seashell that was not merely a relic of a sea creature, but a wind instrument purposefully shaped by the Magdalenian people – early Europeans at the twilight of the last Ice Age. This conch, capable of producing deep and resonant sounds that once might have filled the cave, stands as a testament to our ancestors' ingenuity, transforming simple shells into complex tools that vibrated with the rhythms of their world.

The modifications made to the conch are revealing of the Magdalenian's sophisticated understanding of their materials. Advanced imaging techniques showed that the shell's narrow end was precisely cut to create a small opening, into which a cylindrical mouthpiece – likely a hollow bird bone—was inserted. The edges of the shell's opening were finely sculpted, and traces of resin or wax around the aperture suggest the use of an ancient adhesive, likely to secure the mouthpiece. The shell's exterior was also decorated with ochre-red pigment, mirroring the style of the cave's paintings, and further blending artistry with acoustic innovation.

A skilled musicologist, part of the research team, employed a modern metal mouthpiece to explore the acoustic capabilities of the conch, producing notes close to C, C sharp, and D. This exploration not only confirmed the shell's functionality as a musical instrument but also highlighted the Magdalenian's mastery over sound vibration and resonance. Carbon dating places the conch around 18,000 years old, making it the oldest known human-made conch shell horn and a unique artefact of the European Upper Palaeolithic era, illustrating an advanced level of technological and cultural development.

—

Fritz, C., Tosello, G., Fleury, G., Kasarherou, E., Walter, P., Duranthon, F., Gaillard, P. and Tardieu, J., 2021. First record of the sound produced by the oldest Upper Paleolithic seashell horn. *Science Advances*, 7(7), Article eabe9510. https://doi.org/10.1126/sciadv.abe9510

The poem

Ancient shell sounds

Abandoned at the mouth of your shelter
you quivered apprehensively at our approach,
crying out to be held as we proclaimed
the exception of your discovery.
Sighing wearily as we consigned
you to the dusty silence of our archives.

But now

When I hold you in my hands, I see the face
of your purposefully speckled complexion.
When I lift you to my ear, I hear the sound
of an ancient sea lapping at your shores.
When I place you at my lips, I feel the
heartbeat of your creator pulsing to my breath.

I close my eyes,
as you call out to all
that you have lost.

DIGITALLY UNFOLDING RENAISSANCE LETTERS

The science

In a silent exchange across time, secrets penned long ago have a chance to be heard once more. These letters, once sealed by the crafty hands of their senders, have sat quietly in the folds of history, waiting. The art of keeping messages safe through complex folding, or 'letterlocking', was the encryption of an older world.

Now, with a blend of art and science, these folded treasures are being gently coaxed open, not by hands, but by beams of light too soft to touch yet powerful enough to see through layers of history. X-ray microtomography, a method that lets us look inside objects in fine detail, has been combined with computer algorithms to read these letters without ever breaking their seals. The high sensitivity of the X-ray scanner (designed for intricately mapping the mineral content of teeth) makes it possible to resolve certain types of ink in paper and parchment. Following the X-ray scanning of the letters, computational algorithms were applied to the scan images to identify and separate the different layers of the folded letter and virtually unfold it.

This technique has been used to carefully reveal hidden contents from precious letters written during the Renaissance in Europe, a period that spanned from the 14th to the 17th century and marked a profound revival of art, science, and literature across the continent. These digital tools, powered by algorithms that have learnt from studying a vast collection of letters, are creating a new understanding of historical communication patterns. Voices from centuries ago are now revealed not through physical manipulation, but by using scientific principles to understand how light interacts with different materials.

—

Dambrogio, J., Ghassaei, A., Smith, D.S., Jackson, H., Demaine, M.L., Davis, G., Mills, D., Ahrendt, R., Akkerman, N., Van der Linden, D. and Demaine, E.D., 2021. Unlocking history through automated virtual unfolding of sealed documents imaged by X-ray microtomography. *Nature Communications*, *12*(1), Article 1184. https://doi.org/10.1038/s41467-021-21326-w

The poem

Unlocking letters

Reams of dead letters
hide correspondence
beneath purposeful
cuts and folds –
the contents of
their written past
locked tight
behind the paper-thin
veneer of this
faded piggy bank.

Fold
                          Tuck
          Slit
              Floss
                              Roll

And then adhere.
    Waxn
        Shear
            Route
                Starch
Meanings disappear.

Radiation probes
delicately across the surface,
creating cross structures
to remind us of the pains
taken on behalf
of senders now revealed –
digitally unfurling
history's creases
to open secrets
with seals intact.

MEASURING WHALE SHARKS WITH RADIATION

The science

In the vast ocean, whale sharks glide as the largest fish on Earth, stretching up to 12 metres in length, like submarines patrolling the underwater world. These gentle giants carry the secrets of longevity within them, and scientists have long grappled with the puzzle of their lifespan. The key lies within the whale sharks' vertebrae, which display growth bands akin to the yearly rings of a tree. But interpreting these bands has been a challenge: do they signify the passing of a year or six months?

Scientists have turned to the remnants of the Cold War's nuclear tests for answers. The detonation of nuclear bombs caused a significant spike in atmospheric carbon-14, a radioactive isotope that gradually settled into the ocean's depths and became part of the marine food web. This influx of carbon-14 left a timestamp in every creature, including the elusive whale sharks.

By measuring the carbon-14 embedded in the bands of whale shark vertebrae, scientists have unlocked the growth rings' tempo. Each band faithfully records one year of life, offering a reliable measure of age and growth in these colossal creatures. This radiocarbon method, born of an era marked by atomic anxiety, now serves as a peaceful tool for conservation, shedding light on the whale sharks' true age, which may be as high as 130 years.

Armed with this new understanding, researchers can now chart the life history of whale sharks with greater precision. The bands of time etched in their bones speak of a life quietly woven into the fabric of the oceans – a life that now, finally, can be read with clarity and certainty.

—

Ong, J.J., Meekan, M.G., Hsu, H.H., Fanning, L.P. and Campana, S.E., 2020. Annual bands in vertebrae validated by bomb radiocarbon assays provide estimates of age and growth of whale sharks. *Frontiers in Marine Science*, 7, Article 188. https://doi.org/10.3389/fmars.2020.00188

The poem

Testing the age of sharks

Ageless giants glide beneath the waves,
their stunted snouts sagging coyly
at the impertinence of our horology.
Careless in its senility, one swims
too close to shore –
shallow waters
now shallow grave,
with aged secrets ripe for reaping.
Peeling back the skin
reveal rivulets of cartilage
that glisten like honey –
we count rings as pairs and singlets,
counting off their years with
chronometric precision.
The scars of our past lie buried
in each vertebra,
percolating through the brine –
these cold, neglected sins
leave grubby fingerprints
that fade with time.
Buried memories revealing more
than we could every wish
to see.

MAKING MATERIALS FROM SILK AND CELLULOSE

The science

In crafting the future's materials, scientists seek to blend strength and toughness, a feat that is difficult to achieve. Think of a porcelain plate: strong enough to hold a feast, yet a mere tumble away from shattering. Or a tennis ball: it will withstand a fall but deform under pressure. Bridging this gap requires a delicate balance, one that nature itself has mastered in silken webs and sturdy trees.

Researchers have now tapped into this natural genius, merging the resilience found in spider silk with the steadfastness of plant cellulose. This study marks a leap forward, crafting a composite that stands strong against force and yet resists fracture.

The secret to combining these natural elements lies in a smart design. Scientists use what is called a recombinant protein, which means they have taken genetic information from spider silk and used it to create a new, similar type of protein. This lab-made protein echoes the design of spider silk, known for its incredible ability to stretch and absorb forces without breaking. They match this with cellulose, the tough material found in the walls of plant cells that gives them their rigidity. When these two are paired, the silk-like proteins provide flexibility, allowing the material to absorb impacts, while the cellulose fibres give it a solid structure that resists being squashed or pulled apart. By combining these two elements, they create a new material that is greater than the sum of its parts, with both flexibility and solidity.

These innovative materials offer more than just extraordinary strength and flexibility; they also present a more environmentally friendly option. Traditional plastics, which tend to persist in the environment and create pollution, are tough to get rid of once they have served their purpose. However, these new materials, made from biological substances, have the advantage of being biodegradable. This means they can decompose naturally over time, blending back into the Earth without leaving harmful residues behind. It is a step towards a world where the materials we use no longer linger past their usefulness, reducing pollution and leading us to a more sustainable future.

—

Mohammadi, P., Aranko, A.S., Landowski, C.P., Ikkala, O., Jaudzems, K., Wagermaier, W. and Linder, M.B., 2019. Biomimetic composites with enhanced toughening using silk-inspired triblock proteins and aligned nanocellulose reinforcements. *Science Advances*, 5(9), Article eaaw2541. https://doi.org/10.1126/sciadv.aaw2541

The poem

Natural building blocks

In hollowed holes of
knotted cells,
tensile echoes
bubble beneath the surface –
frothing with potential
as they wait to break
free from the confines
of their cellular walls.
Synthetic solutions
nurtured in nature.

Tangled webs of
dew-lined geometries
glimmer gently,
their native silks
bearing broken bodies
that belay their
precious cargo betwixt
fragile manacles
of suppurating steel.
Synthetic solutions
nurtured in nature.

In empty clean rooms
dexterous fingers replicate
acidic glands,
weaving matrices of artificial silk
through interlocking fibrils;
small and slender fibres
that stitch together proteins
to form structural assemblies
that blend into the worlds
from which they were taken.
Synthetic solutions:
nurtured in nature.

SEEING INTO THE ULTRAVIOLET

The science

Our world is home to creatures with senses far beyond human capabilities. Turtles navigate the vast oceans by sensing Earth's magnetic field, while mantis shrimp perceive the world through polarised light. Elephants communicate with low-frequency rumbles inaudible to human ears, and butterflies see a spectrum extending into ultraviolet (UV) light, which is normally invisible to us. These extraordinary sensory abilities inspire not just awe, but also technological innovation.

Scientists have drawn inspiration from the UV vision of the Chinese yellow swallowtail butterfly to develop a new type of camera. Traditional cameras fail to capture UV light, but this bio-inspired camera is designed to do so. It uses perovskite nanocrystals, which are tiny crystals with special properties that can convert UV light into visible light, allowing the camera to detect it. Additionally, the camera includes stacked photodiodes, which are devices that can detect both visible and UV light and turn these light signals into electrical signals. This innovative design allows the camera to capture a wider range of information than traditional cameras.

This innovation has applications in medicine and beyond, allowing for the accurate identification of biological materials such as amino acids, as well as distinguishing cancerous cells from normal ones. By bridging the gap between the sensory world of butterflies and modern technology, this camera could revolutionise both medical diagnostics and our understanding of the natural world.

—

Chen, C., Wang, Z., Wu, J., Deng, Z., Zhang, T., Zhu, Z., Jin, Y., Lew, B., Srivastava, I., Liang, Z. and Nie, S., 2023. Bioinspired, vertically stacked, and perovskite nanocrystal–enhanced CMOS imaging sensors for resolving UV spectral signatures. *Science Advances*, 9(44), Article eadk3860. https://doi.org/10.1126/sciadv.adk3860

The poem

The butterfly's gaze

The turtle swims
the open seas,
her shell

a compass needle –
magnetic whispers
guide her way,
True North
a gentle force,
of hidden paths
and steady hands.

The mantis shrimp
scans the coral reef
in beams of
splintered light,
beyond the reach
of our mind's eye –
a vibrant mosaic
within the ocean's
deep embrace.

Butterflies flit
on painted wings,
a hidden ultraviolet script
in daylight's light caress.
Pixel-captured dreams
revealing cells,
and life,
and death.
Their world unveiled
and caught –
transforming unseen truths
to snatch the light
once lost.

HOW TREE SNAILS USE SUNLIGHT TO SURVIVE

The science

The Society Islands of the South Pacific, part of French Polynesia, were once a haven for tree snails, with 61 different species thriving across the archipelago. However, the introduction of new species and flawed land management drastically changed this ecosystem.

The giant African land snail, initially brought in as a potential food source, escaped into the wild, decimating local crops. In an attempt to curb this invasive species, authorities introduced the rosy wolf snail, known for its predatory nature, to control the giant African snails.

Unfortunately, the rosy wolf snail turned its appetite towards the indigenous tree snails, decimating their populations. Yet one species, *Partula hyalina*, largely survived the onslaught, raising questions about its resilience.

Researchers proposed that the white shell of *Partula hyalina* enabled it to thrive at the forest fringes, where it could bask in sunlight that other snails, including the rosy wolf snail, could not tolerate. To test this solar refuge theory, scientists used the Michigan Micro Mote, the world's smallest computer, to monitor snail behaviour. These tiny computers were glued directly onto rosy wolf snails and placed on top of and beneath leaves where the *Partula hyalina* snails rested (these snails are a protected species, and so could not have the computers directly affixed to their shells).

By measuring how long each computer took to recharge using solar power, researchers found that at midday, *Partula hyalina* received ten times more sunlight than the rosy wolf snails. This suggests that rosy wolf snails avoid the forest edges, even at night, to prevent being trapped in the Sun before they can retreat.

This research not only explains the relative survival of *Partula hyalina*, but also demonstrates how miniature computers can aid in conservation and land management, highlighting a new frontier for technological solutions in ecological research.

—

Bick, C.S., Lee, I., Coote, T., Haponski, A.E., Blaauw, D. and Foighil, D.Ó., 2021. Millimeter-sized smart sensors reveal that a solar refuge protects tree snail Partula hyalina from extirpation. *Communications Biology*, 4(1), Article 744. https://doi.org/10.1038/s42003-021-02124-y

The poem

Snail survivors

A wolf in snail's clothing
patrols the penumbra,
tentacles twitching
at the dappling of light.

Bathed in sunbeams
your mantle's edge
begins to glow,
a brilliant white
reflecting the certitude
of your asylum.

Losing patience
the wolf retreats,
sliding back
to the darkness
from which we came.

USING SATELLITES TO HELP AFRICAN ELEPHANTS

The science

The African elephant population has plummeted over the past century due to poaching and habitat fragmentation, leaving the species endangered. Monitoring is crucial to conserve and protect these animals, and satellites offer an effective way to count them, avoiding disturbance to both the elephants and data collectors. This also simplifies tracking elephants as they traverse borders, as satellites can orbit without being hindered by national controls or conflict.

However, despite their size, African elephants pose a challenge to satellite detection due to their varied habitats, ranging from shrublands and forests to grasslands. This mix of land types complicates visual identification, even for high-resolution satellite imagery.

Researchers have now utilised imagery from the Worldview-3 and 4 satellites, combined with an automated system that employs machine learning, to detect and count African elephants. This approach has proven to be as accurate as human observers in varied landscapes and nearly as effective in more uniform environments. This demonstrates the potential of using remote sensing and machine learning as practical techniques for wildlife surveying and conservation, leveraging the power of high-resolution satellite imagery.

—

Duporge, I., Isupova, O., Reece, S., Macdonald, D.W. and Wang, T., 2021. Using very-high-resolution satellite imagery and deep learning to detect and count African elephants in heterogeneous landscapes. *Remote Sensing in Ecology and Conservation*, 7(3), pp. 369–381. https://doi.org/10.1002/rse2.195

The poem

Elephants from space

Mechanical eyes
drift across horizons,
capturing hidden
choreographies on
the patchwork cloth
of your vast abode.
Grey on green
on green on grey,
your pixelated torsos
dance fleetingly over
annotated screens
of sharpened reds
and artificial jades.
You flicker from
habitat to habitat –
shrubland to
woodland to
thicket to
plain.
We train our eyes
to count
where you have been,
and imagine what
you might become.

HOW INSECT SWARMS CHARGE THE ATMOSPHERE

The science

The Earth's atmosphere is constantly charged, with its electric fields influenced by precipitation, pollution, volcanism, and even earthquakes. These electric fields can affect weather patterns, influence lightning formation, and impact communication systems. It has now been suggested that clusters of flying insects, which, like many other species, carry a small electric charge, might also play a significant role in atmospheric electricity.

Researchers have measured the electric fields near swarming honeybees and found that such swarms can generate as much atmospheric electric charge as a thunderstorm cloud. This discovery highlights how migrating insects can transport charge in the lower atmosphere, contributing to variations in atmospheric electricity.

These findings have significant physical and biological implications. For example, current climate models do not account for this form of electricity. Additionally, insects are not the only living charge carriers in the atmosphere; birds and microorganisms also carry electric charge. This research therefore underscores the need for further interdisciplinary study into the dynamic interactions between physical and biological entities in the atmosphere.

—

Hunting, E.R., O'Reilly, L.J., Harrison, R.G., Manser, K., England, S.J., Harris, B.H. and Robert, D., 2022. Observed electric charge of insect swarms and their contribution to atmospheric electricity. *Iscience*, 25(11), Article 105241. https://doi.org/10.1016/j.isci.2022.105241

The poem

Buzzing electricity

Hidden fields drift
across the sky,
crowds of life
that crackle with intent
across unsuspecting
tones of milky blues
and scheming greys.
Swarming,
swerving,
flitting,
flowing,
shifting,
thronging,
dancing,

growing.
Living charges
prancing through
the firmament,
with the graceful ease
of the great unknown.

MEASURING EARTHQUAKES WITH DEEP-SEA FISH

The science

In Japan, many people believe that sightings of deep-sea fish in shallow waters signal an imminent earthquake. This folklore has persisted for generations, suggesting that the appearance of such fish might be harnessed for disaster mitigation. In light of this belief, Japanese newspapers have reported sightings of rare deep-sea fish, including oarfish, ribbonfish, dealfish, and unicorn crestfish, providing a valuable dataset for scientific investigation.

Researchers studied over 80 years of data on deep-sea fish sightings and earthquakes in the seas surrounding Japan to test this hypothesis. They found only one potentially correlated event, debunking the folklore and demonstrating that there is no significant connection between deep-sea fish appearances and earthquakes. This result indicates that sightings of these marine creatures do not predict seismic activity and are not useful for future disaster mitigation efforts.

This research not only dispels a cultural myth but also reinforces the importance of evidence-based approaches to disaster preparedness. As the Japanese islands lie along the Ring of Fire, a region prone to seismic activity, accurate and reliable monitoring is essential. While deep-sea fish sightings may continue to captivate the imagination, effective disaster preparedness will rely on scientific monitoring systems such as seismographs and early warning networks, which offer a more dependable foundation for mitigating the impacts of earthquakes.

—

Orihara, Y., Kamogawa, M., Noda, Y. and Nagao, T., 2019. Is Japanese folklore concerning deep-sea fish appearance a real precursor of earthquakes? *Bulletin of the Seismological Society of America*, 109(4), pp. 1556–1562. https://doi.org/10.1785/0120190014

The poem

Deep-sea tremors

From buried cracks
and open wounds
shrouded scales
begin to seep.
Cutting across murky waters
into buried memories
and forgotten half-truths –
inky fingerprints
preserve their presence
alongside court proceedings
and local weather reports.
Over time these sightings
are recalled as prophecies,
portents of the restless Earth
that secrete themselves
as cherished traditions.

Curious custodians
challenge accepted lore,
their digital excavations
laying bare
the absence
of fact.
The fallacy of this folklore
cast beneath the waves –
an empty carcass on which
the ribbonfish can feast.

IMAGING THE EARTH'S SUBSURFACE WITH WHALE SONGS

The science

The most common method for imaging geological structures below
the ocean floor involves releasing loud pulses, or 'shots,' of air
under high pressure from airguns, and then monitoring how these
pulses reflect off the ocean floor. The variations in the reflected

sound waves provide information about the structures beneath. Unfortunately, these loud and repetitive sounds can negatively impact marine animals, damaging their hearing, altering vocalisations that affect feeding, mating, or navigation, and displacing them from their habitats. A more sustainable approach, however, may involve monitoring the songs of fin whales.

Fin whales, the second-largest species on Earth after blue whales, produce low-frequency vocalisations that can be heard up to 1,000 kilometres away, making them one of the most powerful biological sounds in the ocean. Ocean-bottom measurement stations, originally designed to monitor earthquakes, often pick up these vocalisations. Researchers have previously used these recordings to track fin whale movements, but for the first time, these calls are now being used to study the Earth's structure.

The vocalisations of fin whales resemble a series of clicks that can last for hours. Part of the energy from these clicks travels through the ocean, penetrating the oceanic crust, where it reflects and refracts off subsurface structures. This data can then reveal information about the Earth's geological layers, similar to the pulses from airguns.

Though this method produces lower-resolution images compared to airguns, it is significantly less invasive and less expensive. This makes it a complementary approach to conventional studies, offering an environmentally friendly way to explore the geological makeup beneath the ocean's surface.

—

Kuna, V.M. and Nábělek, J.L., 2021. Seismic crustal imaging using fin whale songs. *Science*, 371(6530), pp.731–735. https://doi.org/10.1126/science.abf3962

The poem

Seismic songs

An accidental echo
on the line
oscillates
with the baritone
of your misplaced song.

Waves beneath waves
traverse wires crossed
with the rising beat
of the conductor's
chevron baton.

The reflected signals
of your chorus
reverberate throughout
the deep,
exchanging tones
with every passing

crescendo

as your harmonies
silence our ordnance
with their resonance.

Chapter 3

Environmental physics

INTRODUCTION

In 'Environmental Physics', we explore the complex relationships between physical processes and the environment. We begin with the Madden-Julian Oscillation and its impact on global weather patterns, followed by an examination of how climate change affects cloud formation and rainfall intensity. These shifts underscore the urgent need to understand and respond to our changing climate.

Technological advances such as machine learning provide new insights into predicting extreme weather. Simultaneously, rising sea levels transform coastal forests into ghost forests, highlighting the vulnerability of ecosystems. The rapid melting of Alpine glaciers further underscores the potentially lethal impact of global warming on water sources, emphasising the urgent need for action.

The pervasive nature of plastic pollution is another critical issue in our environment, evidenced by the discovery of nanoplastics in the Alps, posing significant health risks in even these most remote locations. Our exploration also covers the environmental fallout from fireworks, revealing how these celebratory displays leave behind chemical residues that pollute soil, water, and air, causing long-term harm to both wildlife and human health.

On the subject of fallout, the poems in this chapter also explore the long-term impacts of nuclear testing in the Marshall Islands, where persistent radiation continues to affect local communities and ecosystems decades later. This serves as a poignant example of how human actions can leave enduring scars on the environment. Similarly, the recent activity of Tungurahua volcano in Ecuador highlights the ongoing geological threats posed by active volcanoes, with the potential for devastating landslides and eruptions that threaten nearby populations. These examples underscore the persistent and evolving challenges we face in managing and mitigating natural and human-induced environmental hazards.

The physics and poetry featured here further discuss how artificial light emissions disrupt natural night-time environments, affecting both human health and wildlife by altering natural rhythms and behaviours. Meanwhile, zombie fires in the Arctic, which re-emerge from smouldering peat, illustrate the changing nature of fire regimes in a warming world, particularly in remote and sensitive regions.

Environmental Physics invites you to engage with the physical principles that underpin our environmental challenges. It encourages a deeper appreciation of the forces that shape our world, from the macro scale of climate systems to the micro scale of pollutants. Each poem serves as a window into the complex interplay of natural and human influences, offering insights into the urgent need for sustainable practices and thoughtful conservation. Through this lens, the poems aim to inspire a greater understanding and respect for the environmental systems that sustain us.

SHIFTING WEATHER PATTERNS

The science

The Madden-Julian Oscillation (MJO) is a band of rain clouds that moves eastward across the tropical Indo-Pacific region. These clouds travel thousands of kilometres, influencing weather patterns like cyclones, monsoons, and other extreme events across many continents. The MJO's behaviour is influenced by ocean temperatures, particularly in the Indo-Pacific, meaning that climate change-driven warming in this area could significantly affect global weather.

Warming has been occurring throughout the Indo-Pacific, but it is most pronounced over the west Pacific Ocean. This temperature difference drives moisture from the Indian Ocean, enhancing cloud formation. As a result, the MJO's lifecycle has shifted, leading to changing weather patterns worldwide. This includes increased rainfall over southeast Asia and the Amazon, and intensified drying over the west coast of the United States and Ecuador.

This interplay between heat, moisture, and atmospheric dynamics underscores the complex connections between oceanic conditions and global weather patterns. Understanding these relationships is a necessary step for predicting future climate impacts and preparing for changes in weather extremes.

—

Roxy, M.K., Dasgupta, P., McPhaden, M.J., Suematsu, T., Zhang, C. and Kim, D., 2019. Twofold expansion of the Indo-Pacific warm pool warps the MJO life cycle. *Nature*, *575*(7784), pp. 647–651. https://doi.org/10.1038/s41586-019-1764-4

The poem

Warming clouds

Born from synthetic
lines on constructed maps,
clouds rise like anvils
of cotton candy –
a scattering of light
creating apparitions
across the sky.
They drift over tranquil
seas and raging waves,
inhabiting continents –
moving with the unseen
breath of their creator.
Peripheral forces tug
at the edges,
shaping transient forms
with a warming touch
that expands
and contracts
without care.
Patterns shift,
reverberating
through the cloudscape –
thunder shakes the firmament,
while drought stalks
the land below.
Reflected dreams
shimmer

on the surface –
abstract shapes
foretelling
what is yet
to come.

THE FUTURE OF RAINFALL

The science

Around the equator, heavy rainfalls frequently lead to flooding, impacting millions of people and causing significant damage to homes and infrastructure. As the planet warms, it becomes increasingly important to understand how these extreme rain events might change. Predicting the behaviour of intense rainstorms in a warmer world presents challenges, as traditional methods often fail to capture the intricate processes of storm formation and dissipation.

To address this, scientists have made use of detailed computer simulations alongside real-world data to study changes in the heaviest daily rains in tropical regions. Their findings reveal that as the climate warms, rainstorms are not only becoming more frequent but also more intense. Warmer temperatures enable the atmosphere to hold more moisture. This is because warmer air can contain more water vapour. When this moisture eventually condenses into rain, it results in more powerful and heavy rainstorms.

This increased heat energy affects atmospheric dynamics, which are the movements and behaviours of air masses within the atmosphere. For example, warmer air can rise more quickly and carry more moisture. When this air cools and condenses, it releases energy in the form of heat, which further intensifies the storm. This process leads to stronger and more frequent rainstorms.

The organisation of these storms, i.e., how they form and develop, plays a role in their heightened intensity. For instance, storms can become more organised and structured, making them last longer and produce more rainfall. These insights are needed for anticipating future climate impacts and developing improved strategies for managing floods. By understanding these changes, we can better prepare for the challenges posed by more intense and frequent rainstorms in a warming world.

—

Bao, J., Stevens, B., Kluft, L. and Muller, C., 2024. Intensification of daily tropical precipitation extremes from more organized convection. *Science Advances*, *10*(8), Article eadj6801. https://doi.org/10.1126/sciadv.adj6801

The poem

When the sky gathers tears

Clouds gather
like ancient gods
converging,
conspiring,
shedding
secrets of
the coming floods.
Twilight's blankets
cluster,
tightening ranks
over belted waves
to seal their pact
in vaulted skies.
This bloated
congregation
now flails with
righteous pain,
carving sodden welts
through fraying skin.
As the earth
drowns in tears
it never thought
to shed.

MACHINE LEARNING PREDICTS EXTREME WEATHER

The science

Climate change has increased the frequency and intensity of extreme precipitation in recent years. Warmer temperatures lead to heavier rain and snowfall because a warmer atmosphere can hold more moisture. However, pinpointing the specific impacts of climate change on extreme weather events, especially at the regional level, is challenging. Global climate models often lack the spatial detail needed for such precise predictions. Machine learning, which involves computer algorithms that improve through experience, offers a potential solution due to its ability to learn complex patterns at high resolution.

Machine learning algorithms have now been trained to identify atmospheric circulation patterns linked to extreme precipitation in the Upper Mississippi Watershed and the eastern portion of the Missouri Watershed in the United States. This flood-prone area, spanning parts of nine states, has experienced more frequent extreme precipitation and major flooding in recent decades. The algorithm successfully identified over 90% of extreme precipitation days in this region from 1981 to 2019, outperforming traditional statistical methods.

The findings revealed that multiple factors contribute to the increase in extreme precipitation in the Midwest. For instance, since the early 2000s, the atmospheric pressure patterns that lead to extreme precipitation in the Midwest have become more frequent, increasing by about one additional day per year.

By using machine learning in this way, scientists can analyse vast amounts of data to detect subtle patterns in atmospheric conditions that lead to extreme weather. This technology allows for more precise predictions of when and where extreme precipitation might occur, which will help in preparing for and mitigating the impacts of these events. Understanding these changes helps communities better plan for floods and other weather-related disasters, enhancing our ability to respond to the challenges posed by a changing climate.

—

Davenport, F.V. and Diffenbaugh, N.S., 2021. Using machine learning to analyze physical causes of climate change: A case study of US Midwest extreme precipitation. *Geophysical Research Letters*, 48(15), Article e2021GL093787. https://doi.org/10.1029/2021GL093787

The poem

Artificial weather

Buoyant skies linger overhead,
bulging at the seams

with rising menace –
erratic threats
that fall indiscriminately
against the statistical fortitude
of our modelled routines.
In search of clarity,
we train machines
to find patterns
within the cyclical nature
of extreme intent.
Cutting through complexities,
algorithmic digits
point nervously
to the faultless correlation
with our own excess.

RISING SEAS TRANSFORM COASTAL FORESTS

The science

Climate change is rapidly reshaping coastal regions, where ecosystems are especially vulnerable to rising sea levels, increasing salinity, and extreme weather events. As sea levels rise, saltwater moves inland, elevating water tables and salinity levels, which puts stress on freshwater forested wetlands and causes widespread tree mortality. The resulting landscapes, known as ghost forests, are characterised by stands of dead trees and fallen trunks. These changes significantly affect the carbon cycle, as forests that once absorbed large amounts of carbon are replaced by less carbon-dense shrublands and marshes.

The combination of rising sea levels and extreme events like hurricanes and droughts accelerates these ecological shifts. During droughts, reduced freshwater flow allows saltwater to penetrate further inland. Hurricanes contribute by driving storm surges that flood forests with saline water. This interplay creates a feedback loop, where each event worsens the salinisation process. This dynamic is evident in North Carolina's coastal wildlife refuge, where a combination of Hurricane Irene and a prolonged drought led to a rapid increase in ghost forest formation between 2011 and 2012.

Remote sensing technology, which uses satellite or aerial imagery to monitor Earth's surface, has enabled detailed mapping and analysis of these changes over a 35-year period. This technology has revealed that 32% of this refuge's landscape has changed, with significant portions of forest turning into ghost forests and eventually into shrubland or marsh. The patterns show that areas closest to the coast and at lower elevations are most affected due to increased salinity and sea-level rise. These observations highlight the urgent need for conservation strategies to address both the immediate impacts of extreme events and the long-term effects of rising sea levels to protect these ecosystems and their role in the global carbon cycle.

—

Ury, E.A., Yang, X., Wright, J.P. and Bernhardt, E.S., 2021. Rapid deforestation of a coastal landscape driven by sea-level rise and extreme events. *Ecological applications*, *31*(5), Article e02339. https://doi.org/10.1002/eap.2339

The poem

Ghost forest

Surging seas and weeping waves
advance along your coast,
probing buried channels as they
break through the shoreface
to drag briny fingerprints
across weathered limbs
that recoil at the touch.

Tainted tides swell
with pickled poison
as saline sap pours down
your brackish bark,
below a crown of mottled grey
that withers in the drink.

Whisps of memories
linger in brine,
haunting shades
of lost and shattered greens.

METHANE EMISSIONS MOVE WITH THE TIDE

The science

Methane emissions from the Arctic seabed, particularly from gas hydrate reservoirs, are a significant concern due to their potential impact on global greenhouse gas levels. These emissions are influenced by various factors, including ocean temperature increases and sea-level rise, but understanding remains limited, especially in deep-sea environments where data collection is challenging.

Environmental changes in the Arctic, such as tides and sea-level variations, affect methane emissions. Continuous measurements of pore-pressure (the pressure within the sediment pores) and temperature over four days along the west-Svalbard margin show that tides significantly influence the intensity and periodicity (regular intervals) of these emissions. High tides seem to reduce the height and volume of gas emissions, suggesting that rising sea levels might partially counterbalance the potential increase in emissions caused by warming ocean temperatures.

Small changes in sea level can impact methane emissions from the seabed. Lower sea levels during low tides reduce the pressure on sediment pores, causing gas to expand and be released into the water column. This effect was observed to be more pronounced at certain sites, where multiple gas emission events coincided with low tides, emphasising the role of tidal cycles in controlling methane seepage.

Continuous monitoring and combining hydro-acoustic surveys (using sound waves to map underwater features) with in-situ measurements (taken directly at the site) are crucial for accurately assessing methane emissions. Understanding these dynamics is essential for predicting the impact of future sea-level rise and temperature changes on Arctic methane emissions, which has significant implications for global climate models and greenhouse gas inventories.

Sultan, N., Plaza-Faverola, A., Vadakkepuliyambatta, S., Buenz, S. and Knies, J., 2020. Impact of tides and sea-level on deep-sea Arctic methane emissions. *Nature Communications*, 11(1), Article 5087. https://doi.org/10.1038/s41467-020-18899-3

The poem

Lunar flow

Ancient forces dredge secrets
from beneath the seabed,
undulating stimuli
unearthing memories
of a long-buried past.

Rising and falling
with the passing tide
dark omens threaten
to break free
from subterranean
prison cells –
shifting pressures
blossoming cracks
across crumbling,
ill-fitting barricades.

We hold our breath
and pray that the
waters will hold them.

ALPINE GLACIERS MELTING FASTER THAN EXPECTED

The science

Predicting how glaciers will change in the short term is very challenging but essential for understanding water resources, natural hazards, and ecological impacts. Glaciers are dynamic systems where ice flows and melts in response to various factors such as temperature, snowfall, and the underlying terrain. As such, traditional glacier models often face difficulties because they rely on simplified assumptions and struggle to account for the complex interactions between these factors, leading to errors in predictions.

The Instructed Glacier Model (IGM) offers a new approach to these challenges. This model uses advanced deep-learning techniques to process extensive data on ice thickness, how fast the ice moves (surface velocity), and changes in elevation. By integrating this data, the IGM can simulate the behaviour of glaciers with high accuracy and speed. For instance, when tested on the Great Aletsch Glacier, the largest glacier in the Alps, the IGM closely matched observed changes in ice volume over the past two decades. This represents a significant improvement over previous methods.

The IGM's predictions highlight a worrying trend. Even if there is no further increase in global temperatures, Alpine glaciers are expected to lose 34% of their ice by 2050. If the current warming trends continue, this ice loss could rise to 65%. Such rapid ice loss has several serious implications. Glaciers are important sources of fresh water, and their melting reduces the availability of this vital resource. Additionally, the loss of ice increases the risk of glacier lake outburst floods, which occur when the meltwater forms lakes that can suddenly burst their banks, causing destructive floods downstream. The retreat of glaciers also impacts tourism, as many visitors come to see these magnificent ice formations.

—

Cook, S.J., Jouvet, G., Millan, R., Rabatel, A., Zekollari, H. and Dussaillant, I., 2023. Committed ice loss in the European Alps until 2050 using a deep-learning-aided 3D ice-flow model with data assimilation. *Geophysical Research Letters*, 50(23), Article e2023GL105029. https://doi.org/10.1029/2023GL105029

The poem

Retreating futures

In the cold heart
of the Alps,
silent sentinels
unfurl their future
in broken
frozen layers –
their icy tongues
laced
in loss.
Each glacier
a story of time,
etched in crystal
and snow

beneath the dusty cloak
of shifting worlds.
Uncovered prophecies
echo through
empty
valleys –
rivers of ice
flowing towards
tomorrow.
As every fading peak
hums their farewell
to the steady
face of time.

HEAT AND VIOLENCE IN A WARMING WORLD

The science

Firearms have become the leading cause of death for children and adolescents in the United States, with a significant proportion of these deaths resulting from homicides. For every fatal firearm assault, two others survive with injuries requiring emergency care, leading to substantial emotional, physical, and economic burdens. Understanding the factors that contribute to violent incidents is vital, including the potential influence of rising temperatures due to climate change.

Researchers have found a consistent relationship between higher temperatures and an increased risk of shootings in 100 of the most populated cities in the United States. By analysing data from the Gun Violence Archive, which includes more than 116,000 shootings from 2015 to 2020, and correlating it with daily temperature records, they discovered that nearly 7% of all shootings were linked to above-average temperatures. This amounts to almost 8,000 shootings directly associated with higher temperatures.

As temperatures rise, the body's ability to dissipate heat becomes less efficient, leading to an increase in core body temperature. This can cause heat stress, which affects the brain's ability to regulate emotions and control impulses. The increased kinetic energy of air molecules at higher temperatures amplifies the sensation of heat, adding to physical discomfort and agitation.

These rising temperatures also influence social dynamics. On hotter days, more people are likely to be outside, increasing the chances of interactions that could escalate into conflicts. Urban heat islands – areas where urban development causes higher temperatures than in surrounding rural areas – exacerbate this problem. Concrete, asphalt, and buildings absorb and retain heat, raising temperatures in cities. This effect not only increases general discomfort but also heightens the risks associated with heat stress.

As climate change continues to push temperatures higher, these observations underscore the urgent need for policies that help communities adapt to heat while also addressing the risk of heat-related violence. This could include measures to reduce urban heat islands, as well as increasing awareness of the links between heat and aggression.

—

Lyons, V.H., Gause, E.L., Spangler, K.R., Wellenius, G.A. and Jay, J., 2022. Analysis of daily ambient temperature and firearm violence in 100 US cities. *JAMA network open*, 5(12), Article e2247207. https://doi.org/10.1001/jamanetworkopen.2022.47207

The poem

A lethal climate

The soil bursts into flame,
mercury rising
through fevered trends
to bring another kind
of heat –
a frenzied force
that shoots
to maim
and kill.
Collars itching with intent
as triggered fingers
expose fault lines
in how we choose to live –
degrees of harm
unduly falling
on those

already branded
by our febrile,
fatal touch.

NANOPARTICLES IN THE ALPS

The science

Plastic pollution is a widespread issue affecting water, soil, and air across the globe. Over time, plastics break down from larger pieces into microplastics and eventually into even smaller particles called nanoplastics, which are less than one micrometre in size. These tiny particles are especially concerning because they can easily be transported by air, have unique chemical properties, and pose significant biological risks. They can penetrate cell membranes and cause toxicity. Despite these concerns, data on the presence and concentration of nanoplastics in remote natural environments have been scarce.

Researchers conducted a detailed study at the high-altitude Sonnblick Observatory in the Austrian Alps to measure nanoplastic concentrations in surface snow. The study revealed that the average concentration of nanoplastics in the melted snow was 46.5 nanograms per millilitre (ng/mL). To give a sense of scale, one nanogram is one-billionth of a gram, so 46.5 nanograms is an incredibly small amount, but significant given the widespread distribution of these particles.

The most common types of nanoplastics found were polypropylene (PP) and polyethylene terephthalate (PET). Polypropylene is a type of plastic commonly used in packaging and containers, while polyethylene terephthalate is used in products like beverage bottles and synthetic fibres. The researchers discovered that PET concentrations were significantly higher during dry sampling periods. This suggests that during dry weather, nanoplastic particles are more likely to settle onto snow surfaces directly from the air, a process known as dry deposition. This contrasts with wet deposition, where particles are washed out of the atmosphere by precipitation.

Finding nanoplastics in such a remote location has significant implications. It underscores the vast reach of plastic pollution and raises concerns about the potential health risks posed by these particles, even in extremely remote

—

Materić, D., Ludewig, E., Brunner, D., Röckmann, T. and Holzinger, R., 2021. Nanoplastics transport to the remote, high-altitude Alps. *Environmental Pollution*, *288*, Article 117697. https://doi.org/10.1016/j. envpol.2021.117697

The poem

Plastic snow

At the top of the world
you sparkle with seclusion,
sheathed in winter's blade
from the grubby tracks
of those tainted,
foul machines.
But something hidden
lurks beneath,
your flawless lustre
gently glazed
by films of filth
and trade
and greed.
Invisible dust that
falls like a blizzard
across the purity
of your pristine skin,
marking you forever
by the ruined nature
of our see-through sin.

THE HIDDEN COST OF FIREWORKS

The science

Fireworks are a beloved spectacle worldwide, often marking celebrations with vibrant displays of light and sound. However, their impact on the environment and wildlife is far from celebratory. The explosions and vivid colours of fireworks come from rapid chemical reactions and the excitation of metal atoms. When fireworks are ignited, these reactions release gases that expand quickly, producing the explosive sounds and brilliant bursts of light we enjoy. The different colours are generated by metal salts, like strontium for red, barium for green, and copper for blue.

Beyond the spectacle, fireworks leave behind chemical residues that pollute soil, water, and air. These pollutants include heavy metals and perchlorates, which can disrupt thyroid function and accumulate in the environment, posing long-term risks to wildlife and human health. The dispersion of these pollutants is influenced by atmospheric conditions, with wind patterns carrying the particles far from their original site, highlighting the interconnectedness of our environment.

Modern alternatives like 'eco-friendly' fireworks and drone light shows offer promising solutions, providing the same visual spectacle without the associated environmental damage. Drones, for instance, offer a reusable, emission-free option that reduces noise pollution and chemical residues. As awareness of the ecological impacts of fireworks grows, adopting more sustainable practices ensures that our celebrations do not come at the cost of environmental and wildlife health.

—

Bateman, P.W., Gilson, L.N. and Bradshaw, P., 2023. Not just a flash in the pan: short and long term impacts of fireworks on the environment. *Pacific Conservation Biology*, 29(5), pp. 396–401. https://doi.org/10.1071/PC22040

The poem

Ejaculations of excess

A symphony of sparks
echoes through the sky,
lighting up the night
with flaming buds
that hiss
and pop
and roar.
Blooming jewels
whose light and weight
lie heavily
beneath the fading glow

of altered,
after thoughts.
Sensational star-flames
whose stinging grind
lingers long beyond the
flashes of our narrow,
premature delights.

RADIATION EXPOSURE IN THE NORTHERN MARSHALL ISLANDS

The science

Between 1946 and 1958, the United States conducted 67 nuclear tests in the Marshall Islands, releasing significant amounts of radioactive material into the environment. This radioactive fallout has had lasting impacts on both the environment and local populations.

In the Marshall Islands today, measurements show varying levels of gamma radiation, a high-energy form of electromagnetic radiation, that can penetrate most materials and poses a serious health risk to living organisms. While some islands have low radiation levels, others exhibit much higher levels. Soil samples from these islands reveal high concentrations of radioactive isotopes like americium-241, caesium-137, and plutonium. These isotopes, byproducts of nuclear reactions, remain hazardous for decades due to their long half-lives.

The highest radiation levels are often found in the interior regions of some islands, likely due to environmental factors such as runoff, which can concentrate radionuclides (unstable atoms that emit radiation as they decay) in certain areas. For instance, radioactive particles can be washed from higher ground to lower areas during rainfall, leading to higher contamination in valleys and other low-lying areas. The presence of these radionuclides in soil samples from locations such as Runit and Enjebi islands, Bikini, and Naen indicates that these areas remain highly contaminated. The contamination levels often exceed safety limits set for human health, posing risks for resettlement and local food safety.

These observations highlight the ongoing environmental and health challenges faced by the Marshallese people. Understanding the spread and persistence of nuclear fallout contributes to the broader knowledge of how long-term radioactive contamination impacts ecosystems and human communities, reinforcing the need for resilience and careful management of affected areas.

—

Abella, M.K., Molina, M.R., Nikolić-Hughes, I., Hughes, E.W. and Ruderman, M.A., 2019. Background gamma radiation and soil activity measurements in the northern Marshall Islands. *Proceedings of the National Academy of Sciences*, 116(31), pp. 15425–15434. https://doi.org/10.1073/pnas.1903421116

The poem

Contaminated land

Submerged volcanoes
rise up from the ocean floor,
their tips littering the landscape
as errors of the past
cast a barely visible film of filth
across this false pacific paradise.
Where Spanish conquistadors
once prowled the dunes,
scintillators now sweep across the sands,
combing beaches for dirty treasures
that can no longer lie buried.
Pantry islands not suitable for
habitation lie in wait,
their rotten fruit a siren's call
to scattered islanders who cross
impossible craters and
invisible barriers
of radiological taboo.
Scatter plots and stick charts
unshrouding mysteries that
malformed coconuts and
irradiated pandanus
have long known:
that this land
can no longer
be a home.

BULGING VOLCANOES

The science

Tungurahua, the towering 'Black Giant' of Ecuador, has recently shown alarming signs of activity, raising concerns about the stability of its slopes. In late 2015, the west side of Tungurahua swelled rapidly, with the ground shifting up to 3.5 centimetres in just three weeks. This happened alongside more frequent volcanic explosions and earthquakes. The swelling occurred within a large scar left by a massive landslide 3,000 years ago, underscoring the volcano's unpredictable nature.

Advanced computer models have shown that the swelling is mainly caused by magma, molten rock beneath the Earth's surface, being stored in a shallow chamber below the west side of the volcano. The pressure from this magma pushes against the surrounding rock, causing it to bulge outward and creating stress that weakens the volcano's structure. This interaction between the magma and the volcanic rock explains how the movement of molten rock can deform the Earth's surface and potentially destabilise the slope.

If the pressure continues to build, it could trigger another landslide similar to the catastrophic event 3,000 years ago. By studying how magma moves and how it affects the stability of the volcano, scientists aim to predict and manage the risks posed by Tungurahua, ensuring the safety of the communities living nearby.

—

Hickey, J., Lloyd, R., Biggs, J., Arnold, D., Mothes, P. and Muller, C., 2020. Rapid localized flank inflation and implications for potential slope instability at Tungurahua volcano, Ecuador. *Earth and Planetary Science Letters*, *534*, Article 116104. https://doi.org/10.1016/j.epsl.2020.116104

The poem

The Black Giant's collapse

The Earth has set the sky ablaze
with glorious hues of orange.
Strombolian explosions spew
incandescent blocks, emerging

from the Throat of Fire with
biblical intent as they race
down fragile, bulging flanks.

A forgotten scar starts to throb,
its suppurating skin pulsating
with intemperate rage
as beneath its surface a
fresh wound
starts to fester.

Seismic swarms and plumes
of ash threaten to break
free from their swollen,
rounded prison.
The deformed shroud
falling hard
across this spoiled,
expectant land.

THE RISE OF NOCTURNAL POWER EMISSIONS

The science

Artificial light at night, detected via satellites, comes from several sources. These include direct emissions from unshielded outdoor lighting, such as streetlights and building lights, as well as reflected light from surfaces like roads and buildings. Additionally, some light is scattered within the atmosphere. This mix of lights indicates urbanisation, industrial activity, and economic development. However, it also signifies light pollution, which is a growing concern for astronomers and increasingly recognised as a threat to public health and natural ecosystems.

Light pollution refers to the excessive or misdirected artificial light that brightens the night sky, interfering with natural darkness. This phenomenon has several negative impacts. For humans, it has been linked to higher risks of cancer, particularly breast and prostate cancer, due to disrupted circadian rhythms. In wildlife, light pollution leads to declines in populations, changes in ecological communities, and disruptions in essential ecosystem services like pollination. For instance, many nocturnal animals rely on natural darkness for hunting and navigation, and artificial light can disorient them, leading to decreased survival rates.

From 1992 to 2017, global satellite-detectable light emissions surged by at least 49%. However, this estimate might be conservative. The transition to LED (Light Emitting Diode) technology, which emits more light at wavelengths not easily detected by current satellites, suggests that the actual increase in visible radiance could be much higher—potentially up to 270% globally, and as much as 400% in specific regions. LEDs, while more energy-efficient and longer-lasting than traditional lighting, emit more blue light, which scatters more in the atmosphere and contributes significantly to skyglow, the brightening of the night sky over inhabited areas.

Understanding the sources and impacts of artificial light is crucial for developing strategies to mitigate light pollution. This includes designing better lighting systems that minimise unnecessary light emissions and implementing policies to reduce light pollution, thereby protecting both human health and the environment.

—

Sánchez de Miguel, A., Bennie, J., Rosenfeld, E., Dzurjak, S. and Gaston, K.J., 2021. First estimation of global trends in nocturnal power emissions reveals acceleration of light pollution. *Remote Sensing*, 13(16), Article 3311. https://doi.org/10.3390/rs13163311

The poem

Dirty light

The Earth tries to sleep,
casting off the shadows
of a distant star
beneath the tattered veil
of greying night.
Behind thinning eyelids
the atmosphere erupts,
burning with the embers
of ferocious solid states.
Cloaked in filthy lustre,
these irradiating irritations
avoid detection

from ageing, weary eyes.
Ever-present emitters
in the clear light
of never-ending day.

ARCTIC FIRES THAT COME BACK TO LIFE

The science

In early 2020, the Arctic experienced unprecedented wildfires, which started two months earlier than usual. A significant factor in these wildfires was the re-emergence of underground smouldering fires, known as 'zombie fires'. These fires persist in carbon-rich peat and reignite when conditions become favourable. The severity and early onset of the 2020 fires have raised concerns about a potential shift in the Arctic's fire regime.

Zombie fires, which can burn slowly underground throughout the winter, pose a unique challenge. When temperatures rise and conditions dry out, these fires can reignite and spread. The 2020 Arctic fires highlighted the need to understand these persistent fires and their greenhouse gas emissions. Extreme temperatures and drying conditions are making landscapes that were previously resistant to fires more vulnerable. Over half of the fires detected in 2020 occurred on ice-rich permafrost, which is ground that remains frozen for two or more consecutive years. The heat from the fires accelerates the thawing of permafrost, releasing stored carbon into the atmosphere and further exacerbating climate change.

Current models for predicting biomass burning, which is the burning of organic matter such as trees and plants, are inadequate for the unique conditions in the Arctic. The fires in 2020 underscored the need for new tools to differentiate between holdover fires (those that persist from previous seasons) and new ignitions. Effective monitoring requires on-the-ground observations to assess changes in surface elevation and water levels. Collaborating with Indigenous and local communities is crucial, as they provide valuable long-term observations and local knowledge that can guide research efforts.

To address this global challenge, a coordinated international effort is necessary. Support from forums like the Arctic Council can help develop a comprehensive Arctic fire monitoring system that integrates traditional knowledge with modern scientific techniques.

This approach will improve our ability to predict, monitor, and manage wildfires in the Arctic, ultimately helping to mitigate their impact on the environment and climate.

—

McCarty, J.L., Smith, T.E. and Turetsky, M.R., 2020. Arctic fires re-emerging. *Nature Geoscience*, *13*(10), pp. 658–660. https://doi.org/10.1038/s41561-020-00645-5

The poem

Zombie fires

Buried beneath the snow line,
these smouldering corpses
begin to glow.
Forgotten fires,
whose reanimated embers
burn brightly
across the tundra;
frozen bodies recoiling
at the heat
of their undying embrace.

Thawed to life by
distant warming
these undead hordes now
straddle horizons;
crimson fingers
flickering over
blue-veined memories,
as they dance
impossibly beyond
the water's edge.

Nervously
we shift our gaze
towards a restless earth,
as unwanted resurrections
blaze across the landscape.

REVIVING FORGOTTEN FIRE MANAGEMENT

The science

Recent research into the Monte Pisano region of Italy reveals how historical agropastoral practices, such as litter raking and managed burning, effectively reduced the flammability of Mediterranean landscapes. These practices, largely carried out by women, have been understudied and marginalised due to historic state prohibitions and a lack of scientific interest. This abandonment has significantly contributed to increased fire risks, as forests have become overgrown and laden with combustible material, exacerbated further by rising summer temperatures and droughts due to climate change.

Despite the stigma around managed burning, which has often been illegal and thus ignored by officials, these traditional methods played an important role maintaining less fire-prone landscapes. Likewise, traditional land-use practices, particularly those carried out by marginalised groups like women, need to be recognised and revitalised to help mitigate current wildfire challenges. For instance, litter raking, which removed substantial biomass and reduced flammability, was an effective practice whose ecological impacts have been overlooked due to its gendered nature and nonmarket status.

By documenting the role of these traditional practices, we can advocate for their integration into modern fire prevention and landscape management strategies. This approach acknowledges the historical and cultural significance of these practices while also leveraging their practical benefits in reducing fire risk in the Mediterranean and potentially other regions experiencing similar issues due to land abandonment and climate change.

—

Mathews, A.S. and Malfatti, F., 2024. Wildfires as legacies of agropastoral abandonment: Gendered litter raking and managed burning as historic fire prevention practices in the Monte Pisano of Italy. *Ambio*, *53*, pp. 1065–1076. https://doi.org/10.1007/s13280-024-01993-x

The poem

Women of the wildfires

In Monte Pisano's shade
women tread,

silent stewards
to the leaf
and flame.
Their hands rough
as the bark they skirt,
a patchwork quilt
for earth's unmade bed.
Beneath their fingers
memories stir,
murmuring
through groves,
through vines –
every unturned blade
a sunken spark.
They move like ghosts
amongst the trees,
their legacy
a faded collage
of presence
and loss.
And still
the setting stands,
voiceless witness
to its quiet keepers –
a ritual unspoken
a fire tamed
a disaster unbirthed.

Chapter 4

Celestial physics

INTRODUCTION

In this, our third and final theme of poetry in this book, we turn our gaze towards the stars, with 'Celestial Physics' exploring the vast and intricate dance of the cosmos. The poems are ordered in such a way that invites us to start from our Sun and move outwards, a very heliocentric view of our universe, which has been shown to be scientifically inaccurate, yet offers a structured and familiar narrative to our exploration.

Our journey begins with the Sun, our closest star, whose dynamic and often violent behaviour has been studied to understand its impact on our solar system. Through these explorations, we listen in on the quiet periods of solar activity, uncovering the subtle signals that reveal much about its baseline behaviour. This foundational understanding sets the stage for predicting solar phenomena that affect space weather and, consequently, life on Earth.

As we move beyond our Sun, we encounter the remnants of cosmic events that have shaped planetary bodies and entire ecosystems. The narrative takes us back thousands of years to ancient cities destroyed by powerful cosmic airbursts, linking historical events with celestial phenomena. This intertwines the destructive and creative forces of the universe, highlighting how cosmic occurrences can influence the course of human history.

The exploration continues with the mysteries of cosmic dust, where we unravel the origins of extraterrestrial objects that offer insights into the materials and processes that form the building blocks of planets. By examining these unique compositions, we gain a glimpse into the interstellar origins of cosmic dust, enriching our understanding of the diversity and complexity of matter in the universe.

DOI: 10.1201/9781003514688-4

Further, we extend our reach to the search for life beyond Earth, where the spectra of microorganisms found in Earth's icy environments provide a reference for detecting life on distant, icy exoplanets. This quest underscores the interconnectedness of all life forms and the continuous pursuit of knowledge that drives human curiosity and exploration.

Towards the end, our focus shifts to the grand architecture of the universe itself. The mapping of the cosmic web, inspired by the network-building abilities of slime mould, illustrates how galaxies are interconnected through vast, invisible structures. This innovative approach offers a clearer understanding of the universe's large-scale structure, linking theoretical predictions with actual observations.

In this thematic collection, each poem serves as a bridge between the scientific and the poetic, offering a reflective lens through which we can appreciate the grandeur and mystery of the cosmos. From the familiar warmth of our Sun to the farthest reaches of the universe, 'Celestial Physics' invites us to consider our place and purpose in the vast expanse of space and time.

EAVESDROPPING ON THE SUN

The science

The Sun, our closest star, is a dynamic and complex object. While its most dramatic outbursts, like solar flares and sunspots, often capture our attention, the Sun is never completely still. Even during periods of minimal activity, subtle processes continue to occur on its surface and in its atmosphere.

During these quiet periods, the Sun's subtle radio signals provide valuable insights. Observations from radio observatories, such as the Metsähovi Radio Observatory in Finland, have focused on particularly low activity phases. For instance, during a phase between May and September 2019, faint radio emissions were examined even when no noticeable active regions were present on the solar surface.

These weak radio signals are mainly linked to two features on the Sun: coronal holes and magnetic bright points. Coronal holes are areas where the Sun's surface is less dense, emitting fast solar winds as streams of charged particles flow into space. Magnetic bright points are small, bright spots on the Sun's surface, often possessing structures called flux tubes, which are channels that guide magnetic energy.

By comparing the radio data with extreme ultraviolet observations, the sources of these radio signals can be identified. Bright points with flux tubes show stronger emissions, indicating more intense magnetic activity. In contrast, bright points without flux tubes have weaker signals, suggesting less activity. Monitoring these subtle signals during quiet periods will lead to improvements in the predictions of solar activity that can impact space weather and communication systems on Earth.

—

Kallunki, J., Tornikoski, M. and Björklund, I., 2020. Identifying 8 mm radio brightenings during the solar activity minimum. *Solar Physics*, 295(7), Article 105. https://doi.org/10.1007/s11207-020-01673-5

The poem

Our quiet star

We trace your violence
with methodic unease,
charting chaotic ferocities
as measured outbursts –
a humanised hoax
to retain control.

As your temper subsides
we cut loose the charts,
averting our eyes to
let you gently slumber –
a brokered peace
to catch breath.

You twitch restlessly
in brazen sleep,
excitable forces conspiring
to escape your surface –
passive brightenings
of unknown intent.

Your sudden flare
catches us unprepared,
we consult our graphs
to correct their errata –
minimal signs
hidden from view.

THE COSMIC DEVASTATION OF TALL EL-HAMMAM

The science

Around 1650 BCE, approximately 3,600 years ago, a dramatic event destroyed the ancient city of Tall el-Hammam in the southern Jordan Valley, northeast of the Dead Sea. This catastrophic event was caused by a cosmic airburst, an explosion in the atmosphere from a meteor or comet fragment, which was even more powerful than the well-known 1908 Tunguska event in Russia. The Tunguska explosion involved a space object around 50 metres wide, releasing energy a thousand times greater than the Hiroshima atomic bomb.

Evidence from Tall el-Hammam includes a thick, city-wide layer rich in carbon and ash, filled with materials subjected to extreme conditions. This layer contains shocked quartz, indicating pressures of 5–10 gigapascals, melted pottery and mudbricks, and various high-temperature minerals such as diamond-like carbon, soot, and metallic spherules (small, spherical particles that can form through various natural processes). Some of these spherules include elements like iron, silicon, calcium carbonate from melted plaster, and even melted platinum, iridium, nickel, gold, silver, zircon, chromite, and quartz. These findings suggest that temperatures during the event exceeded 2,000 degrees Celsius.

The airburst's impact was so intense that it demolished the city's palace complex, which stood 4–5 stories tall, and destroyed the massive mudbrick rampart that was 4 metres thick. Human remains found in the area show extreme dislocation and fragmentation, indicative of a violent, high-energy event.

In addition to the immediate physical destruction, the airburst also brought a significant amount of salt to the area, making the soil hypersaline and unsuitable for agriculture. This environmental change caused the abandonment of Tall el-Hammam and about 120 other settlements within a 25-kilometre radius for 300–600 years.

—

Bunch, T.E., LeCompte, M.A., Adedeji, A.V., Wittke, J.H., Burleigh, T.D., Hermes, R.E., Mooney, C., Batchelor, D., Wolbach, W.S., Kathan, J. and Kletetschka, G., 2021. A Tunguska sized airburst destroyed Tall el-Hammam a Middle Bronze Age city in the Jordan Valley near the Dead Sea. *Scientific Reports*, *11*(1), pp. 1–64. https://doi.org/10.1038/s41598-021-97778-3

The poem

An ancient burst of salty air

Muddied bricks and
melted sheets of bronze
disclose the
revelation of
your erasure.
Between molten pots
and shattered bones
a flash of heat
scars the earth,
fractured branding
imbued with the debris
of your obliterated past.
Fire,
brimstone,
air,
and salt,
flicker in silent witness
that you were ever
really there.

THE ORIGINS OF COSMIC DUST

The science

The Hypatia stone is a small stone found in Egypt in 1996, named after Hypatia of Alexandria, who was a prominent female philosopher, astronomer, mathematician, and inventor of the late 4th and early 5th centuries.

Using proton-induced X-ray emission (PIXE) spectroscopy, Hypatia's elemental composition has been analysed, revealing significant differences from solar system objects. PIXE is a technique that uses a proton beam to determine the elements present in a sample, and the stone has been found to contain two different types of material: one almost purely carbon and another rich in heavier elements like iron but lacking in silicon. This pattern suggests that Hypatia's parent body formed from interstellar dust (found in space between stars), rather than within our solar system.

Hypatia's unusual elemental makeup, especially its low silicon content, aligns with theoretical models of supernova Ia explosions. These supernovae occur when a white dwarf star, a small, dense remnant of a star, explodes due to accumulating matter from a companion star. The intense nuclear reactions during this explosion break down heavier elements into lighter ones, creating a specific pattern of element distribution. This connection implies that the stone's origin involves a supernova, highlighting the complex processes that contribute to the formation of cosmic dust.

The study of Hypatia offers insights into the variety of interstellar dust and the formation of materials in our solar system. Understanding these processes helps explain the diversity of matter in the cosmos and provides a clearer picture of the early solar nebula's formation, i.e., the cloud of gas and dust that eventually formed the Sun and planets of our solar system.

—

Kramers, J.D., Belyanin, G.A., Przybyłowicz, W.J., Winkler, H. and Andreoli, M.A., 2022. The chemistry of the extraterrestrial carbonaceous stone "Hypatia": A perspective on dust heterogeneity in interstellar space. *Icarus*, *382*, Article 115043. https://doi.org/10.1016/j.icarus.2022.115043

The poem

Supernovae in the stones

Twisted in a spiral of spite,
two stars collide in pain –
shockwaves of remorse
that echo over lengths
once impossible
to fathom.
Seared in desperate rage
these remnants of conflict

gather dust
in the void
of our betweens.
Scattered over burning sands,
the violence of their birth
etched into faces
that we trace
with trembling hands.

A CATALOGUE OF LIFE IN ICE

The science

The search for life in the universe is entering an exciting phase with the discovery of thousands of exoplanets and new missions exploring our solar system. To identify life on icy exoplanets and moons, reference data from Earth's icy environments is essential. A spectra catalogue of life in ice has been developed, measuring the reflection spectra of 80 microorganisms from ice and water, collected from locations such as Hudson Bay and Great Whale River in Kuujjuarapik, Quebec. This catalogue, covering visible to near-infrared wavelengths, serves as a guide for detecting life signs on icy worlds, highlighting carotenoid pigments as significant indicators.

Carotenoids, pigments found in many life forms, play roles in photosynthesis and protection against environmental stresses. On Earth, these pigments create vibrant colours in microorganisms found in icy environments. By studying the reflection spectra of these organisms, unique signatures that indicate life can be identified. Dried samples of these microorganisms show even higher reflectance, suggesting that life signs might be more detectable on exoplanets and moons with dry surfaces compared to those with abundant liquid water.

Reflectance spectra reveal how light interacts with the surface of these microorganisms. When light hits the surface, some wavelengths are absorbed while others are reflected, creating a spectral signature unique to each pigment. By comparing these Earth-based signatures to observations from telescopes, scientists can search for similar patterns on distant icy worlds. This method focuses on the physical properties of light reflection and absorption to identify potential biosignatures.

This spectra catalogue will enable scientists to search for life beyond Earth by providing a reference for identifying biological pigments on icy exoplanets and moons. As we prepare for upcoming missions and the deployment of advanced telescopes in space, this catalogue will play an important role in our quest to discover whether we are alone in the cosmos.

—

Coelho, L.F., Madden, J., Kaltenegger, L., Zinder, S., Philpot, W., Esquível, M.G., Canário, J., Costa, R., Vincent, W.F. and Martins, Z., 2022. Color catalogue of life in ice: surface biosignatures on icy worlds. *Astrobiology*, 22(3), pp. 313–321. https://doi.org/10.1089/ast.2021.0008

The poem

Colours of distant life

A vibrancy fans out
across the bay,
blues and greens
tipping over into reds
and cannot-be-seens;
vivid signatures
sketched out in bands
across the frozen,
broken water.
Carefully we pick the hues,
dropping colours
into spectral baskets –
bouquets of ice
to catalogue new life
amongst the stars.

THE LUNAR ANTHROPOCENE

The science

Humans have always been explorers, venturing into new and challenging environments. From our early beginnings in Africa, we have spread across the entire planet, adapting to all sorts of climates and terrains. Our ability to thrive in diverse conditions is a unique aspect of being human. Throughout our history, we have left traces of our presence wherever we go – from ancient tools and artwork to changes in the environment. This impact is so significant that some scientists believe we have started a new geological era, known as the Anthropocene, defined by human influence on Earth.

The exploration of space, particularly the Moon, is the latest chapter in this story of human expansion. The race to the Moon in the mid-20th century marked the beginning of our journey beyond Earth. Just like on Earth, we have left our mark on the Moon. There are now footprints, equipment from lunar missions, and even craters caused by spacecraft. These traces are not just historical artefacts; they represent a significant human impact on an entirely new environment.

As space travel becomes more common, with private companies planning tourism and mining operations on the Moon, it is time to consider if we have started a new era on the Moon – a 'Lunar Anthropocene'. This concept suggests that just as we have deeply affected the Earth, our influence is now extending to other celestial bodies as well.

—

Holcomb, J.A., Mandel, R.D. and Wegmann, K.W., 2024. The case for a lunar anthropocene. *Nature Geoscience*, 17(1), pp. 2–4. https://doi.org/10.1038/s41561-023-01347-4

The poem

Footprints on the Moon

Beneath the sun's
restless stare
we sprung
curious,
boundless,
ravenous.
Our tracks
a tangled web
across the fractured
lines of living canvas.
Tools of flint
and sparks of thought
etched our story
in stone,
and clay,

and bark.
In the furrows of farmed fields.
In the skeletons of skyscrapers.
In the choked escape
of muddied streams.
Unsated
our gaze turned skyward,
to a pearl
cast adrift
amongst the ocean
of stars.

RADAR OBSERVATIONS ON MARS

The science

The Jezero Crater on Mars is an ancient impact crater, thought to have once housed a lake, making it a prime site for studying the planet's geological and hydrological history. NASA's Perseverance rover, equipped with a ground-penetrating radar called RIMFAX, is exploring the western edge of Jezero Crater. This radar provides continuous subsurface images, allowing scientists to see up to 20 metres below the surface.

The data from the RIMFAX ground-penetrating radar revealed a distinct separation between the older crater floor and the younger delta deposits in Jezero Crater. The older crater floor layers are irregular and sloped, showing signs of significant erosion, meaning they have been worn away over time. In contrast, the younger delta layers are flat and horizontal, suggesting they were formed in a calm lake. This boundary, called an unconformity, indicates that the crater floor was eroded before the delta layers were deposited on top of it.

The delta deposits offer valuable clues about Mars' past. The flat, horizontal layers, especially at the base, suggest they formed in a stable, calm lake. Such an environment would have been ideal for preserving signs of ancient life, known as biosignatures. The presence of an unconformity also indicates that Mars experienced significant geological changes, with periods of erosion followed by periods of sediment build-up, showing a dynamic history.

The findings from the RIMFAX radar enhance our understanding of the geological history of Jezero Crater. They confirm that the delta sediments are younger than the crater floor and highlight the complex interplay of erosion and deposition that has shaped this region. These insights will help in interpreting the broader geological and hydrological evolution of Mars and in planning future missions aimed at further uncovering the red planet's secrets.

—

Paige, D.A., Hamran, S.E., Amundsen, H.E., Berger, T., Russell, P., Kakaria, R., Mellon, M.T., Eide, S., Carter, L.M., Casademont, T.M. and Nunes, D.C., 2024. Ground penetrating radar observations of the contact between the western delta and the crater floor of Jezero crater, Mars. *Science Advances*, *10*(4), Article eadi8339. https://doi.org/10.1126/sciadv.adi8339

The poem

Over Martian shores

A riddle wrapped in
burnished dust
turns its
wounded face
towards our gaze –
furrowed brows
guarding tales
of distance,
water,
life.
A blanket of history
stretched out
across this bed
of broken mouths.
Every layer a
story told in pride,
as our envoy's
dead and certain eyes
peer deep
into the withered
surface soul.
Layers
once flat and plane
now tilt
disrupted,

revealing ghosts
of ancient lakes
calm and still
until
a break.
A pause
before the delta's birth,
and within
the trembling
hopeful sounds
of life.

SATURN'S GIANT STORMS

The science

Saturn experiences giant storms approximately every 20–30 years, creating immense cloud disturbances that encircle the entire planet. The most recent storm, in 2010, provided extensive data thanks to the Cassini spacecraft. Observations from the Very Large Array (VLA) in 2015 have revealed that these storms leave deep and enduring effects on Saturn's atmosphere, particularly impacting the distribution of ammonia gas.

The VLA data showed significant changes in ammonia concentrations before and after these giant storms. After the 2010 storm, regions expected to be rich in ammonia were found to be depleted. This depletion is likely due to a process where moist air rises and then dry air descends, effectively stripping the atmosphere of ammonia in those regions. Additionally, the VLA detected lingering effects from previous storms, some of which are hundreds of years old, indicating persistent bands of altered ammonia concentrations.

The 2010 storm exhibited unique behaviour, splitting into two distinct components that moved in opposite directions along Saturn's latitudinal lines, leaving a noticeable gap in ammonia concentration at around 43°N latitude. This split and the resulting gap suggest complex atmospheric dynamics influenced by Saturn's zonal winds, which can create barriers preventing the anomalies from spreading further.

Future observations will provide a more comprehensive understanding of these phenomena. The VLA's 2015 observations focused on Saturn's northern hemisphere, as the rings obscured the southern hemisphere. The next opportunity to observe both hemispheres will be in 2025, when Saturn's rings are edge-on to Earth. These upcoming observations will contribute to further understanding of the long-term effects of giant storms and the differences between the northern and southern hemispheres, shedding light on the unique and complex nature of Saturn's weather systems.

—

Li, C., de Pater, I., Moeckel, C., Sault, R.J., Butler, B., deBoer, D. and Zhang, Z., 2023. Long-lasting, deep effect of Saturn's giant storms. *Science Advances*, 9(32), Article eadg9419. https://doi.org/10.1126/sciadv.adg9419

The poem

The storms of Saturn

A bracelet
of cosmic splendour,
banded clouds
wrapped tight with
swirling stripes of haze
to mask the
rage within.
Tempests prowl
broad rings,
ravenous brutes
whose chaos-breath
engulfs the skies.
Violent beasts
who come
and go
as they please,
their long shadows
still,
creeping,
changing.
Storm-ghosts from centuries past,
whose memory burns bright
in cloud and wind

long after
they are dead
and gone.

FINDING EARTH-SIZED PLANETS

The science

Understanding the early stages of planetary formation is essential to our comprehension of the universe. Scientists use young planets to test theories about how planets form and change over time. In recent research, astronomers made a significant finding: they identified an Earth-sized planet, HD 63433 d, orbiting a young, Sun-like star just 22 light-years away. This discovery was made using data from the Transiting Exoplanet Survey Satellite (TESS) and the Gaia space observatory.

HD 63433 d is part of a star system that also includes two larger planets, all orbiting a star similar to our Sun but much younger, at 414 million years old (in comparison, our Sun is about 4.6 billion years old). This star is part of the Ursa Major moving group, a collection of stars that share a common origin and travel through space together. By studying the light curves from TESS, which show how the star's brightness dips as planets pass in front of it, scientists determined that HD 63433 d is slightly larger than Earth and completes an orbit around its star in just over four days.

This system's proximity to Earth and the brightness of its star make it an excellent target for further study. Observing HD 63433 d will enable scientists to learn more about the processes that strip away or preserve planetary atmospheres, especially in young planets. The findings offer a rare opportunity to study an Earth-sized planet at a critical stage in its development, providing valuable insights into the formation and evolution of other planets, including our own.

—

Capistrant, B.K., Soares-Furtado, M., Vanderburg, A., Jankowski, A., Mann, A.W., Ross, G., Srdoc, G., Hinkel, N.R., Becker, J., Magliano, C. and Limbach, M.A., 2024. TESS Hunt for Young and Maturing Exoplanets (THYME). XI. An Earth-sized Planet Orbiting a Nearby, Solar-like Host in the 400 Myr Ursa Major Moving Group. *The Astronomical Journal*, *167*(2), Article 54. https://doi.org/10.3847/1538-3881/ad1039

The poem

Cosmic reflections

In the whispers of dust
a fledgling orb
entwines
a youthful star –
the genesis of worlds
bound within
each
uncanny spin.
Playing out
the permutations
of how
we might
be born
and live
and die.
In its light
our past
reflected,
our future
foretold.
Two distant
mirrored pieces,
adrift
across the
patchwork quilt
of time.

IN SEARCH OF SUPERHABITABLE PLANETS

The science

The quest to find planets beyond our solar system that might be even better suited for life than Earth has led scientists to consider the concept of 'superhabitable' planets. While Earth is rich with life, it may not represent the pinnacle of habitability in the universe. The idea of superhabitable planets suggests that there are celestial bodies with conditions that could support more biomass and biodiversity than Earth. By focusing on these superhabitable planets, scientists hope to improve the chances of finding extraterrestrial life.

Superhabitable planets are thought to exist within specific astrophysical parameters that make them more conducive to life than Earth. These parameters include factors such as the type of star the planet orbits, the planet's mass, its surface temperature, and its atmospheric composition. For example, planets orbiting K dwarf stars may offer more stable environments for life due to their longer lifespans compared to our Sun, a G dwarf star. Additionally, a planet slightly larger than Earth, with more landmass and shallow water areas, might support more diverse and abundant life forms.

Several key characteristics could make a planet superhabitable. These include a planet being slightly older than Earth, which allows more time for life to develop and evolve. The planet's surface should be warmer by a few degrees to create a more extensive habitable zone. A thick, moist atmosphere rich in oxygen and the presence of a large moon to stabilise the planet's climate are also important factors. Plate tectonics, which recycles nutrients and supports diverse habitats, is another feature for maintaining a superhabitable environment.

Among the thousands of exoplanets discovered, scientists have pinpointed 24 potential superhabitable candidates. These planets meet several of the criteria for superhabitability, such as being in the habitable zone of K dwarf stars and having suitable temperatures and ages. However, due to current technological limitations, we cannot yet confirm all the necessary conditions for these planets. As our observational capabilities improve, particularly with missions like NASA's planned Starshade spacecraft, we may soon be able to identify and study these superhabitable worlds more closely, prioritising them in the search for extraterrestrial life.

—

Schulze-Makuch, D., Heller, R. and Guinan, E., 2020. In search for a planet better than earth: Top contenders for a superhabitable world. *Astrobiology*, 20(12), pp. 1394–1404. https://doi.org/10.1089/ast.2019.2161

The poem

Better than Earth

Mechanical eyes sweep
immeasurable skies,

searching for life with
boundaries pre-defined
by the limitations
of our existence.
Blinkered hands reaching
into dusky bags
to pull blue marbles
from a kaleidoscope
of planetary possibilities.
Not younger or smaller
but warmer and wetter,
these are the states
in which life would be better.
Habitats that turn
our planet green,
as it nervously shifts
explorations
to the watery blocks
upon which
we are stacked.

REFINING STAR DISTANCES WITH ASTEROSEISMOLOGY

The science

Parallax is a method used by astronomers to measure the distance to stars. It involves observing the apparent shift in a star's position against the background of more distant stars as the Earth orbits the Sun. The Gaia spacecraft, operated by the European Space Agency (ESA), uses this technique to create an extensive map of our galaxy, providing detailed information about the positions and movements of stars.

However, Gaia has faced challenges with parallax offsets, which are small errors in the distance measurements of stars, particularly those that are farther away. To address these issues, scientists have turned to a technique called asteroseismology, which studies oscillations within stars. They focused on a group of stars known as red clump stars, which are particularly useful for measuring distances because they have consistent brightness and are relatively distant.

Scientists selected 3,500 red clump stars from a larger group of red-giant stars observed by space missions like Kepler, K2, and TESS. By analysing these stars, they aimed to refine the distance measurements provided by Gaia. They examined various factors that could affect these measurements, such as different methods for analysing star oscillations, the impact of interstellar extinction (dust and gas that dim and redden starlight), and the quality of data from different light analysis surveys. By focusing on the highest quality data and minimising potential errors, they improved the accuracy of Gaia's parallax measurements.

One observation was that for stars dimmer than a certain level, previous correction models were quite accurate. However, for brighter stars, the distance measurement errors varied significantly depending on their location in the sky. This suggests that current models might not fully account for these positional differences, particularly for brighter stars where there are fewer reference points available. The results matched well with previous studies, further validating the methods used and offering a clearer picture of Gaia's measurement accuracy.

—

Khan, S., Miglio, A., Willett, E., Mosser, B., Elsworth, Y.P., Anderson, R.I., Girardi, L., Belkacem, K., Brown, A.G., Cantat-Gaudin, T. and Casagrande, L., 2023. Investigating Gaia EDR3 parallax systematics using asteroseismology of Cool Giant Stars observed by Kepler, K2, and TESS-I. Asteroseismic distances to 12 500 red-giant stars. *Astronomy & Astrophysics*, 677, Article A21. https://doi.org/10.1051/0004-6361/202347919

The poem

Musical measurements of the spheres

In the cosmic depths,
where silence swells
a billion beacons
drift
against the ocean
of the night.
Spread across
the endless dark,

this stellar symphony
chart their course –
each astral pulse
a map to distant shores,
amongst the
flailing dance
of dying light.
Starry arias,
whose waves yield
mass,
and range,
and sight.
Celestial echoes
intoned
from spheres of gold,
their songs
as old as time,
waiting
to be heard.

THE CANNIBALISM OF STARS

The science

In binary star systems, an intriguing phenomenon occurs where one star gradually strips material from its companion, a process known as stellar cannibalism. As this predatory star absorbs matter from its victim, it accelerates in its rotation, eventually evolving into what astronomers term a Be star. These Be stars are distinguished by their rapid spin and the emission of gas discs.

The unique features of Be stars, such as their fast spinning and the special light they emit, are due to movements within the star called nonradial pulsations. These pulsations happen because of the star's high speed. Many Be stars gain their speed from interactions with a nearby companion star in a binary system, where two stars orbit each other closely.

Using advanced methods like interferometry, which involves combining light from multiple telescopes to see details more clearly, and radial velocity measurements, which track how a star moves towards or away from us, astronomers have discovered faint, stripped-down companion stars. These companions can be different types of stars, including helium-burning subdwarfs, white dwarfs, or neutron stars.

By studying the orbits of these companion stars, scientists have gathered valuable information regarding how they evolve and behave over time. This research has greatly improved our knowledge of how stars interact and change within binary systems.

——

Klement, R., Rivinius, T., Gies, D.R., Baade, D., Mérand, A., Monnier, J.D., Schaefer, G.H., Lanthermann, C., Anugu, N., Kraus, S. and Gardner, T., 2024. The CHARA Array interferometric program on the multiplicity of classical Be stars: new detections and orbits of stripped subdwarf companions. *The Astrophysical Journal*, 962(1), Article 70. https://doi.org/10.3847/1538-4357/ad13ec

The poem

Consumed stars

Material drifts as dream,
grand paths etched
into the void
charting distance,
hunger,
light.
Twin hearts
beat as one,
burning dead
and bright
across the stage
of time.
One pulse surges,
its partner
prey,
stripped bare
by gravity's jaw.
The core of the consumed
casting off its skin
in silent rage
to shimmer
in the stellar tide –

fading in the dark
for all it gave
in vain.

HOW NEUTRON STARS INSPIRE SAFER NUCLEAR WASTE

The science

One significant challenge in the adoption of nuclear energy is managing nuclear waste, which remains hazardous for extremely long periods. Transforming these waste products into more stable forms is an ongoing scientific endeavour.

Transmutation involves adding neutrons to unstable elements, effectively reversing the process of nuclear decay. Inspired by observations of gravitational waves from merging neutron stars, this technique aims to make nuclear waste safer by transforming it into heavier, more stable versions of itself. Although direct observation of transmutation is difficult, it can be estimated by studying reactions involving selenium-79 (79Se), a byproduct of nuclear waste.

In the study of neutron capture by 79Se, scientists use a process called the (d, p) reaction. In this process, 79Se is bombarded with deuterons, causing it to emit a proton and capture a neutron, transforming into 80Se. This reaction is analysed using inverse kinematics, where the roles of the projectile and target are reversed, allowing for better detection and analysis of the reaction products.

Advanced tools, such as the RI Beam Factory at the RIKEN Nishina Center in Wako, Japan, measure gamma rays emitted from excited states of selenium isotopes. Gamma rays, which are high-energy photons, provide vital information about the energy levels and reactions occurring within the nucleus. By measuring these emissions, scientists can determine neutron capture cross-sections, which indicate the probability of a neutron being captured by 79Se.

These measurements align well with theoretical models and existing data, confirming the effectiveness of the experimental methods used. Understanding neutron capture in selenium-79 will aid in the development and design of future nuclear waste management facilities and provide insights into the astrophysical phenomena related to the formation of heavy elements.

—

Imai, N., Dozono, M., Michimasa, S., Sumikama, T., Ota, S., Hayakawa, S., Hwang, J.W., Iribe, K., Iwamoto, C., Kawase, S. and Kawata, K., 2024. Neutron capture reaction cross-section of 79Se through the 79Se (d, p) reaction in inverse kinematics. *Physics Letters B*, *850*, Article 138470. https://doi.org/10.1016/j.physletb.2024.138470

The poem

Nuclear transmutations

In the river of stars,
waves crash
and burn –
a dance of particles,
unseen,
unfelt,
unknown.
Neutrons,
like murmurs,
collide in the night,
absorbing the call
to shift,
transform,
assume –
a new weight,
not decay
but growth
reversing the flow
to dampen a storm.
From the heart
of a star
to the soul
of a world.

MAPPING THE COSMIC WEB WITH SLIME

The science

Modern cosmology, the study of the universe's origin and structure, suggests that matter forms a vast network known as the 'cosmic web'. This web consists of thread-like structures connecting different regions of the universe. However, most of this matter is either dark, meaning it does not emit light, or does so too sparsely to be seen directly. While galaxy surveys have revealed some parts of this web, these visible galaxies make up less than 10% of all normal matter, leaving much of the cosmic web unobserved.

Inspired by the remarkable network-building abilities of the slime mould *Physarum polycephalum*, an innovative method has been developed to map the cosmic web. Slime moulds are simple, single-celled organisms that grow and move to find food efficiently. They form complex, vein-like networks to transport nutrients, providing insights into optimised pathways and networks in nature.

A computer model mimicking the slime mould's behaviour was applied to data from galaxy surveys, confirming that most of the intergalactic medium (IGM) – the diffuse gas found between galaxies – resides within these web-like structures. By observing how hydrogen in the IGM absorbs light from distant quasars (extremely bright objects powered by black holes), it was found that hydrogen absorption increases in denser areas of the cosmic web.

This innovative approach offers a clearer understanding of the universe's large-scale structure. It connects theoretical predictions with actual observations, demonstrating how galaxies are linked through these invisible filaments. As this work continues, it will provide deeper insights into how galaxies and their surrounding gas interact, revealing the complex relationships that shape the universe's vast and hidden framework.

—

Burchett, J.N., Elek, O., Tejos, N., Prochaska, J.X., Tripp, T.M., Bordoloi, R. and Forbes, A.G., 2020. Revealing the dark threads of the cosmic web. *The Astrophysical Journal Letters*, 891(2), Article L35. https://doi.org/10.3847/2041-8213/ab700c

The poem

Moulded galaxies

Across dank, shaded
beds of cosmic potential
specks of light flicker
into life.
Emerging from chaos,

filaments of matter
weave their nets –
unclassifiable structures
that stretch
across the emptiness
as ethereal webs that
glisten in the dewy dawn.

Barely trodden paths
glow faintly in the dark,
celestial causeways that
trace the congruence of
our galactic ancestry.
Their hidden alleyways
conspicuous by the
mass of what we
cannot see –
the weight of their
neglect
calling to us
from across the void.

Blindly, we turn our
eyes to the sky,
searching for breadcrumbs
between the pathways –
oblivious to knowledge
that has always been with us:
the map of our universe
rolled tight within a single cell.

UNVEILING HIDDEN GALAXIES

The science

Even with the advances of modern astronomy, some galaxies remain hidden from traditional optical surveys. These galaxies, often shrouded in dust, only become visible in the far-infrared (FIR) spectrum. A particular focus is on galaxies from over 12 billion years ago, thereby giving us a better understanding of the early universe.

A recent discovery highlights one such galaxy, AzTECC71, identified using the James Webb Space Telescope (JWST). This galaxy, previously invisible at wavelengths shorter than 850 micrometres, was detected through its FIR emissions. Located in the COSMOS-Web survey field, AzTECC71 is one of the reddest galaxies identified, which suggests it has a high amount of dust and lies at a great distance. This galaxy is both massive and extremely luminous in the infrared spectrum, indicating it is a significant site of star formation.

What makes AzTECC71 particularly intriguing is its near-invisibility in the near-infrared spectrum, detectable only with the advanced capabilities of JWST. Traditional methods, using telescopes like the Hubble Space Telescope, failed to spot this galaxy. By examining how this galaxy absorbs and emits light, scientists can infer its characteristics, such as its stellar mass and the rate at which new stars are forming.

The study of AzTECC71 provides a vital clue to a larger, elusive population of similar galaxies that could vastly outnumber previously estimated counts. If more FIR-bright, optically-faint galaxies are found, it would suggest that the early universe hosted many more star-forming regions than currently thought. This challenges existing models of galaxy formation and evolution, potentially requiring a reassessment of how galaxies have developed over cosmic time.

—

McKinney, J., Manning, S.M., Cooper, O.R., Long, A.S., Akins, H., Casey, C.M., Faisst, A.L., Franco, M., Hayward, C.C., Lambrides, E. and Magdis, G., 2023. A Near-infrared-faint, Far-infrared-luminous Dusty Galaxy at z~ 5 in COSMOS-Web. *The Astrophysical Journal*, *956*(2), Article 72. https://doi.org/10.3847/1538-4357/acf614

The poem

Twilight galaxies

Hidden deep
in time,
a tapestry
of galaxies
lie blinking

in the gloam,
veiled in youth
beyond
the fading sight
of narrow bands
and prying eyes.
A distant,
fiery blaze
of giants
robed in red,
their burnished cloaks
a muted light
within the creases
of the cosmos.
Spectral isles cast
in dark's embrace,
they drift
and play
like phantoms
in the night.
Stories etched
in starlight's trace,
ghostly glimpses
revealing more
than we could
ever dream
to see.

Chapter 5

Writing physics poems

INTRODUCTION

Welcome to the final chapter in our journey of physics and poetry: writing your own poems. In this chapter, I will guide you through the process of creating your own poetry that captures the essence of scientific discovery and the wonder of the physical world. We will explore how to write science poetry in different forms – haiku, sonnet, and ghazal – and understand how each form can enhance the communication of scientific ideas.

As we have explored throughout this book, poetry offers a unique way to engage with physics. It allows us to explore and express scientific concepts in a manner that transcends traditional academic communication. Poetry, with its rich language and evocative imagery, enables us to tap into the emotional and imaginative aspects of scientific inquiry. This blend of art and science can lead to a deeper appreciation and understanding of the physical world, making complex ideas more accessible and resonant.

In this chapter, we will discuss specific techniques for writing poetry about physics. Additionally, we will look at how rhythm and structure can mirror scientific principles, creating a harmonious balance between form and content. Whether you are new to poetry or have never written anything related to science before, this chapter will provide you with the guidance and confidence needed to start. By the end of this chapter, you will have the tools and inspiration to start writing poetry that reflects the wonders of physics, while speaking to the human experience of discovery and curiosity.

THE PROCESS OF WRITING SCIENCE POETRY

Writing science poetry involves several steps that blend creativity with scientific understanding. This process allows you to transform complex scientific concepts into engaging and evocative poetic expressions. Here is a structured approach that I adopt in my own writing, which should help you get started:

1. **Read abstracts.** Start by reading scientific abstracts or papers to find inspiration and highlighting any words or phrases that stand out. Abstracts provide a concise summary of research and can be a rich source of imagery and themes for your poetry. By distilling the key elements of a study, abstracts highlight the essence of scientific discoveries, offering a fertile ground for poetic exploration.

2. **Find your poetic style.** Identify the poetic style that best suits the subject matter and your personal voice. As will be discussed below, different forms can bring out various aspects of the physics you are communicating. Experiment with different forms to see which one resonates most with the themes you wish to explore.

3. **Draft.** Write your first draft without worrying about perfection. Focus on capturing the essence of the science and how it resonates with you. Let your initial ideas flow freely, allowing the scientific concepts to merge naturally with your poetic voice.

4. **Edit.** Refine your poem, paying attention to both scientific accuracy and poetic expression. Consider feedback from peers (and especially non-scientists) to enhance your work. This step involves fine-tuning the language, making sure that the scientific details are correct, and polishing the imagery to make your poem both informative and aesthetically pleasing.

5. **Rewrite.** Writing is rewriting. Make multiple passes through your poem to improve clarity, imagery, and rhythm. Each revision helps you refine your thoughts and expressions, making the poem more coherent and impactful. Do not be afraid to make significant changes to enhance the poem's overall effect. Make sure to save your earlier drafts, as they might contain lines to use in other poems.

6. **Share.** Share your poetry with others. This could be through readings, publications, or online platforms. Engaging with an

audience can provide valuable feedback and further inspiration. Sharing your work allows you to connect with others who appreciate the fusion of science and poetry and can lead to new insights and collaborations.

Of course, this is only one approach, and while it works for me, it may not work for everyone. Feel free to adapt these steps to suit your own creative process. The key is to find a method that allows you to effectively blend your scientific understanding with your poetic expression, creating works that are both meaningful and impactful.

THE IMPORTANCE OF POETIC FORM

Choosing a specific poetic form can provide a creative scaffold that guides your writing, much like the structure of an experiment guides scientific inquiry. Each form comes with its own set of rules and characteristics that lend rhythm and structure to your poem, but they can also help to convey scientific concepts more effectively. By choosing the right structures, you can find a rhythm and flow that enhances both the artistic and communicative aspects of your writing, allowing you to distil complex ideas into a concise and impactful format.

For example, if you are writing a science poem about measuring the loss of sea ice, a nonet (a nine-line poem that starts with nine syllables in the first line and decreases by one syllable in each subsequent line) can effectively mirror the gradual reduction of ice. The diminishing syllables reflect the shrinking ice, creating a visual and rhythmic representation of the research findings. This structural choice both supports the content of the poem and also adds a layer of meaning through its form, making the scientific message more poignant and memorable.

Understanding the rules of a poetic form is akin to learning the principles of classical mechanics before venturing into the more abstract realm of quantum mechanics. Classical mechanics, with its clear and predictable laws, provides a solid foundation for understanding the physical world. Similarly, traditional poetic forms offer a framework within which you can practice and perfect your craft. Once you are comfortable with these forms, you can experiment with breaking the rules in ways that serve your creative vision, much like how scientists explore beyond classical mechanics to uncover the bizarre and fascinating phenomena of the quantum world.

Just as quantum mechanics reveals the underlying complexities and peculiarities of the universe, breaking the conventional rules of poetry can lead to innovative and profound expressions of scientific ideas. However, this creative freedom is most effective when it is built upon a deep understanding of the traditional forms. By first learning some of the established rules, you gain the tools and confidence to manipulate and transform them, creating poetry that resonates on multiple levels. This balance between structure and creativity allows you to explore and communicate the intricacies of physics in a manner that is both accessible and evocative.

Here I present three poetic forms –the haiku, sonnet, and ghazal – chosen because they represent different cultures, styles, and approaches to conveying scientific concepts. Each form offers unique advantages for expressing scientific ideas through poetry. Let us explore each one in turn, discussing their basic structure, scientific connection, characteristics, and examples from my own practice.

Haiku

Basic structure

The haiku is a poetic form from Japanese literature that traditionally focuses on nature and the seasons. Widely embraced and adapted in various languages and cultures, haiku originally captures the essence of a moment, distilling it into a brief, vivid image or feeling. Western adaptations often follow a pattern of five syllables in the first line, seven in the second, and five in the third. However, this strict syllable count is not necessary because Japanese haiku count on (音), or morae, which are phonetic units shorter than syllables. As a result, the rhythm and structure of Japanese haiku do not translate directly into English syllables, making the precise syllable count less relevant.

Instead, the focus should be on capturing the spirit and brevity of the original form, allowing for flexibility in syllable count to convey the intended image or emotion effectively in English. In modern haiku, the syllable count is often disregarded. The goal is to be concise, creating a snapshot of a single moment or experience. This brevity and focus enable the poet to convey a deep, often poignant, insight with minimal words. A key feature of haiku is the use of juxtaposition, where two contrasting images or ideas are placed next to each other. This technique creates a deeper resonance and invites the reader to find a connection or insight between the seemingly unrelated elements.

Scientific connection

The brevity of haiku forces the poet to distil complex scientific ideas into their essence, highlighting contrasts and connections. The form's juxtaposition mirrors the often surprising and revelatory nature of scientific discoveries. In haiku, the juxtaposition of two images or ideas can create a deeper resonance, reflecting the moment of insight or revelation often found in scientific research.

Characteristics

- **Essence of a moment.** Haiku captures the essence of a moment, focusing on a specific scene or event. This form of poetry excels in distilling a fleeting moment into a few words, creating a vivid snapshot that evokes a sense of immediacy and presence. The goal is to transport the reader directly into the moment, allowing them to experience the sights, sounds, and emotions as if they were there themselves.
- **Nature and human nature.** While haiku often touches on human experiences, these are typically intertwined with natural elements, reflecting the interplay between humans and the environment. This connection highlights the symbiotic relationship between people and nature, offering insights into how our lives are influenced by the natural world. Senryu, in contrast, focuses primarily on human nature and social issues, often using humour or satire to explore everyday human behaviours and societal norms, without the necessity of a natural context.
- **17 syllables or less.** Traditional haiku are written in a 5–7–5 syllable pattern, but modern haiku in English may be shorter to avoid unnecessary words. The emphasis is on brevity and precision, helping to make sure that each word contributes to the overall impact of the poem. This economy of language forces the poet to focus on the most essential elements of the moment they are capturing, often resulting in a more powerful and resonant piece.

- **Two-part structure.** Haiku are typically written in two parts: a fragment (often the first or last line) and a phrase (the remaining two lines). This structure creates a juxtaposition between the two parts, encouraging the reader to find connections and contrasts that deepen the meaning of the poem. The fragment and phrase can capture different aspects of a scene or moment, bringing together seemingly disparate elements in a way that creates a richer, more nuanced picture.
- **Present tense.** Haiku are written in the present, focusing on the here and now rather than past events. This use of the present tense helps to convey a sense of immediacy and engagement, drawing the reader into the moment being described. By focusing on the present, haiku captures the transient beauty of a single instant, emphasising the fleeting nature of our experiences and the importance of mindfulness.
- **Season word (kigo).** Most haiku include a word that indicates a specific season, known as kigo. These words could refer to flora, fauna, or festivals associated with that season. The use of seasonal references helps to anchor the haiku in a specific time and place, adding layers of meaning and enhancing the reader's understanding of the natural setting. Kigo can evoke a wide range of associations and emotions, enriching the poem with cultural and environmental significance.

Example

> Zebrafish tails mend
> waves ripple new buds open –
> Cells in sync regrow.

This haiku is inspired by research on the role of mechanical waves in zebrafish tail regeneration (Matsubayashi, 2023). The study reveals that when a zebrafish's tail is amputated, mechanical waves propagate from the wound edge, helping the tissue to heal and regrow precisely. These waves are driven by hydrogen peroxide (H_2O_2) signals, which vary depending on the amputation position. Closer cuts produce more H_2O_2, leading to stronger waves and faster regeneration. This mechanism highlights the complex interplay between physical and biochemical signals in tissue repair and growth, providing new insights into regenerative biology.

The haiku captures the essence of the scientific discovery by distilling the dynamic healing process into a vivid and immediate image. The reference to 'new buds' serves as a subtle kigo, grounding the poem in the early spring season associated with renewal and growth, while the imagery of waves and cells aligns with the natural and biological themes of the study. The present tense and two-part structure of the haiku create a sense of immediacy and engagement, effectively conveying the transient beauty and precision of the regenerative process.

Sonnet

Basic structure

The sonnet is a form of poetry from Italian literature, popularised by poets like Francis Petrarch and later adopted and adapted by English poets such as William Shakespeare. A sonnet is a poem consisting of 14 lines, traditionally written in iambic pentameter. Iambic pentameter means that each line has ten syllables arranged in a pattern where an unstressed syllable is followed by a stressed syllable (da-DUM, da-DUM). The most common types of sonnets are the Shakespearean and the Petrarchan.

A Shakespearean sonnet has a rhyme scheme of ABABCDCD EFEFGG, where each letter represents the end sound of a line. Lines ending with the same letter rhyme with each other. The poem is structured into three quatrains (four-line sections) followed by a final rhymed couplet (two-line section). This means that the first and third lines of the quatrain rhyme with each other (A), as do the second and fourth lines (B), and this pattern continues with new rhymes for each quatrain (C with C, D with D, and so on). The final two lines (GG) form a rhymed couplet that often delivers a resolution or twist to the poem's theme. Each quatrain typically explores a distinct idea, with the final couplet providing a satisfying conclusion.

Conversely, a Petrarchan sonnet features a rhyme scheme of ABBAABBACDCDCD and is divided into an octave (eight-line section) and a sestet (six-line section). The octave introduces a problem or scenario, while the sestet provides a response or reflection. The shift or turn in the argument, known as the volta, usually occurs at the beginning of the sestet in a Petrarchan sonnet, and in the final couplet of a Shakespearean sonnet. This structure and the use of rhyme and rhythm help to create a musical quality and thematic coherence within the poem.

Scientific connection

The structured form of the sonnet, with its defined rhyme schemes and rhythmic patterns, mirrors the progression of scientific thought – from hypothesis to discovery, or from problem to solution. The volta, or turn, in a sonnet introduces a new perspective or insight, reflecting the moment of revelation often found in scientific research.

Characteristics

- **14 lines.** A sonnet is always comprised of 14 lines written in iambic pentameter. This structure, with its ten-syllable lines following an unstressed-stressed pattern, creates a rhythmic flow that enhances the poem's clarity and emotional impact.
- **Rhyme scheme.** Sonnets follow specific rhyme schemes, which vary by type. The Shakespearean sonnet employs an ABABC DCDEFEFGG pattern, while the Petrarchan sonnet uses ABBA ABBACDCDCD. These patterns provide a musical quality and unify the poem's thematic elements.
- **Volta.** The volta, or 'turn,' is a shift in the poem's argument, perspective, or theme. In a Petrarchan sonnet, it typically appears at the start of the ninth line, whereas in a Shakespearean sonnet, it often occurs in the final couplet. This shift adds depth and complexity, allowing for a nuanced exploration of the theme.
- **Thematic development.** Sonnets often explore a single theme throughout the poem. In the Shakespearean sonnet, each quatrain may address different facets of the theme, with the final couplet providing a resolution. In the Petrarchan form, the octave sets up the theme or problem, and the sestet offers a resolution or reflection.
- **Language and imagery.** The language in sonnets is typically rich and evocative, using metaphors, similes, and vivid imagery to convey complex ideas and emotions. The fixed structure encourages careful word choice, meaning that each line contributes significantly to the poem's overall impact.

Example

In lands where parched earth meets the deluge's wrath,
Dry winds give way to torrents' might and fight.
A cycle fierce that leaves a winding path,
Of cracked, dry soil and rivers in full flight.
The drought, it saps the land of all its life,
Then sudden rains bring floods with ruthless sway.
The balance tipped, disaster's double strife,
As both extremes mark nature's harsh display.
Yet in this dance of dry and wet, we find,
A lesson deep within the earth's embrace.
To understand the shifting of the mind,
And how extremes in nature interlace.
For in this world's great web of cause and plight,
We seek to learn, to guard, and to unite.

This Shakespearian sonnet was inspired by research on the inter-actions between droughts and floods on a global scale, analysing historical occurrences and their impacts on flood severity and timing (Matanó et al., 2024). The study highlights that approximately 24% of floods globally are preceded by or occur during drought conditions, with significant variations based on regional climates. It was found that floods following droughts tend to have a sever-ity distribution similar to single flood events, while those occurring during droughts are generally of lower magnitude, especially in arid regions. The timing of floods can be delayed by drought conditions, affecting early warning systems and disaster risk management. This analysis underscores the need for integrated approaches in man-aging the risks associated with these compound and consecutive events, particularly in the context of a warming climate.

A sonnet, with its structured and rhythmic nature, provides an ideal form to convey the complex interplay between drought and flood events. The 14-line format allows for a focused yet expansive exploration of themes, mirroring the intricate balance of environ-mental forces discussed in the research. Additionally, the use of the turn, or volta, in the ninth line introduces a shift in perspective that deepens the reader's understanding, mirroring the sudden and dramatic transitions between drought and flood, encapsulating the unpredictable and cyclical nature of these phenomena. The Shakespearean sonnet, in particular, was chosen over the Petrarchan form due to its flexible structure of three quatrains followed by a final couplet. This allows for a progressive development of ideas and themes, with each quatrain exploring different aspects of the research before the couplet provides a succinct resolution or insight.

Ghazal

Basic structure

The ghazal is a form of poetry from Arabic literature that has been adopted and adapted by various cultures, including Persian, Urdu, and Hindi literary traditions. A ghazal is a poetic form composed of a series of rhyming couplets, each known as a sher. The ghazal is composed of a minimum of five sher – and typically no more than fifteen – that are structurally, thematically, and emotionally autonomous, meaning that each couplet can stand alone as a complete thought or idea. Despite this independence, all the couplets in a ghazal are linked by a common refrain, called the radif, and a rhyme, known as the qaafiya.

The radif is a repeated phrase or word that appears at the end of each couplet, and which for the first couplet also appears at the end of the first and second line. The qaafiya is an internal rhyme that comes just before the radif in each couplet. This creates a consistent pattern throughout the entire poem, giving it a unique musicality and rhythm. For example, if the radif is 'night' and the qaafiya is a word that rhymes with 'impending', such as 'never-ending', each couplet would end similarly, such as '...impending night', '...never-ending night', and '...mending night'. Therefore, choosing the radif and qaafiya is very important when writing ghazal, as otherwise you will quickly run out of options.

In addition to the internal rhymes and refrain, all lines in a ghazal are written in the same meter. Meter refers to the rhythmic structure of the lines, such as the iambic pentameter discussed above in relation to the sonnet. In a ghazal, this consistent rhythmic pattern is maintained across all lines, providing a uniform cadence to the poem. The uniform meter creates a cohesive auditory experience, binding the individual couplets together through a shared rhythmic foundation, enhancing the overall musicality and flow of the ghazal.

A final distinctive feature of the ghazal is the inclusion of a proper name, often the poet's name or pen name in the final couplet. This personal signature adds a unique touch to the poem, making it a personal expression of the poet's voice and identity.

Scientific connection

The repetitive nature and intimate structure of a ghazal can reflect the continuous pursuit of knowledge and the personal connection between a scientist and their subject. Each couplet, though self-contained, adds to the overarching theme, similar to how individual experiments or studies build upon each other to advance scientific understanding. The use of refrain and rhyme creates a sense of continuity, highlighting the persistent quest for insight and discovery in science. Additionally, the inclusion of the poet's (or another's) name adds a subjective element, introducing an emotional dimension that might be absent in the scientific narrative itself.

Characteristics

- **Couplets.** The ghazal is composed of between five and 15 rhyming couplets, where each couplet (or sher) is a complete thought or statement. The first couplet sets the rhyme and refrain pattern, which the subsequent couplets follow. This couplet structure allows each sher to stand alone while contributing to the overall theme.
- **Refrain and rhyme.** Each couplet ends with the same refrain (radif) and is preceded by an internal rhyme (qaafiya). This creates a musical and rhythmic consistency throughout the poem.
- **Meter.** All lines in a ghazal share the same meter, providing a uniform rhythm. The consistent meter helps to unify the individual couplets into a cohesive whole.
- **Personal touch.** The final couplet often includes a proper name, adding a personal signature to the poem. This tradition personalises the ghazal, making it a unique expression of the poet's voice and identity.
- **Thematic independence and unity.** While each couplet can stand alone, addressing a unique aspect of the theme, they collectively contribute to the overarching message or feeling of the ghazal.

Example

In cosmic dark, faint galaxies arise,
In halos deep, the ancient skies arise.

Dwarf gatherings in shadows subtly form,
From hidden realms, their shapes and size arise.

These implied masses hint at secrets vast,
A universe where unseen ties arise.

As warm dark matter's secrets gently bloom,
In cosmic dance, where unheard cries arise.

Now Euclid's gaze shall pierce through night's disguise,
Revealing truths where hollow eyes arise.

This ghazal is inspired by research into how faint dwarf galaxies form and the dark matter halos that influence their numbers (Nadler et al., 2024). The study uses detailed simulations of the Milky Way to create models of these small galaxies, exploring what we can learn about galaxy formation and dark matter. The findings show that observing all satellite galaxies around one Milky Way-like galaxy can provide some limits on how galaxies form, but it is not always precise. However, combining data from multiple galaxies can reduce uncertainties. The study also projects that observing complete satellite populations can help us understand different types of dark matter, such as warm dark matter, and how it compares to the more commonly known cold dark matter.

The ghazal was chosen for this research because its repetitive nature and structured format mirror the continuous and intricate pursuit of understanding dark matter and galaxy formation. Each couplet, though independent, contributes to the overarching theme, much like individual observations and simulations build a comprehensive picture of cosmic phenomena. The refrain and rhyme create a sense of continuity and persistence, reflecting the ongoing quest for knowledge in astronomy. The inclusion of the pen name 'Euclid' in the final couplet ties the poetic form to the ESA's Euclid mission, which is designed to explore the composition and evolution of the dark Universe.

FINAL WORDS

This book began with the aim of bridging the gap between the accessibility of scientific research and the appreciation of non-specialist readers. By blending scientific inquiry with poetic expression, it sought to demonstrate how the appreciation of one domain can enhance the understanding of the other. Through themes such as the transient beauty of entropy, the ambiguity of quantum mechanics, and the symbolism of light, we explored the complementary nature of physics and poetry.

As we conclude, I hope these poems inspire you in your physics, writing, science communication, and personal learning journeys. Poetry has the unique power to distil complex scientific ideas into accessible and engaging narratives, making it a valuable tool for education and outreach. Let these poems be a catalyst for creativity and understanding, bridging the gap between the empirical and the existential. Together, we can continue to explore and appreciate the richness of the universe through both scientific and poetic lenses.

I also encourage you to write your own physics-themed poems and share them with others, helping to communicate and enrich our collective understanding of the universe and our place within it. If you have any questions, feedback, or would like to share your work, please get in touch. Your contributions and insights are important, and together we can continue to explore the intersection of science and poetry.

Thank you for reading,
Sam

REFERENCES

Matanó, A., Berghuijs, W.R., Mazzoleni, M., de Ruiter, M., Ward, P.J. and Van Loon, A.F., 2024. Compound and consecutive drought-flood events at a global scale. *Environmental Research Letters*, 19(6), Article 064048. https://doi.org/10.1088/1748-9326/ad4b46

Matsubayashi, Y., 2023. Mechanical waves help zebrafish regrow their tails. *Nature Physics*, 19(9), pp. 1241–1242. https://doi.org/10.1038/s41567-023-02151-y

Nadler, E.O., Gluscevic, V., Driskell, T., Wechsler, R.H., Moustakas, L.A., Benson, A. and Mao, Y.-Y., 2024. Forecasts for Galaxy Formation and Dark Matter Constraints from Dwarf Galaxy Surveys. *The Astrophysical Journal*, 967(1), Article 61. https://doi.org/10.3847/1538-4357/ad3bb1

Index